B

Naum Ya. Krupnik

Banach Algebras with Symbol and Singular Integral Operators

Translated from the Russian by A. Iacob

1987

Birkhäuser Verlag
Basel · Boston

Author's Address:
Prof. Naum Yakovlevich Krupnik
Kishinev State University
Sadovaya 60
277003 Kishinev
USSR

Translation of:
Banakhovy Algebry s Simbolom i Singulyarnye
Integral'nye Operatory, Kishinev, Shtiintsa, 1984

For this translation the Russian text was revised by the author.

Library of Congress Cataloging in Publication Data

Krupnik, N. Ya. (Naum Yakovlevich), 1932–
    Banach algebras with symbol and singular integral operators.
    (Operator theory, advances and applications ; vol. 26)
    Rev. translation of: Banakhovy algebry s simvolom i
singuliarnye integral'nye operatory.
    Bibliography: p.
    Includes index.
    1. Banach algebras.   2. Integral operators.
I. Title.   II. Series: Operator theory, advances and
applications ; v. 26.
QA326.K7813   1987   512'.55   87–25005
ISBN 0-8176-1836-8 (U.S.)

CIP-Kurztitelaufnahme der Deutschen Bibliothek

Krupnik, Naum:
Banach algebras with symbol and singular integral
operators / Naum Ya. Krupnik. Transl. from the
Russian by A. Jacob. – Basel ; Boston : Birkhäuser,
1987
    (Operator theory ; Vol. 26)
    Einheitssacht.: Banachovy algebry s simbolom
    i singulyarnye integral'nye operatory
    ‹engl.›
    ISBN 3-7643-1836-8 (Basel)
    ISBN 0-8176-1836-8 (Boston)
NE: GT

© 1987 Birkhäuser Verlag Basel
Printed in Germany
ISBN 3-7643-1836-8
ISBN 0-8176-1836-8

PREFACE

About fifty years ago S.G. Mikhlin, in solving the regularization problem for two-dimensional singular integral operators [56], assigned to each such operator a function which he called a *symbol*, and showed that regularization is possible if the infimum of the modulus of the symbol is positive. Later, the notion of a symbol was extended to multidimensional singular integral operators (of arbitrary dimension) [57, 58, 21, 22]. Subsequently, the synthesis of singular integral, and differential operators [2, 8, 9] led to the theory of pseudodifferential operators [17, 35] (see also [35(1)–35(17)]*), which are naturally characterized by their symbols.

An important role in the construction of symbols for many classes of operators was played by Gelfand's theory of maximal ideals of Banach algebras [20]. Using this theory, criteria were obtained for Fredholmness of one-dimensional singular integral operators with continuous coefficients [34 (42)], Wiener-Hopf operators [37], and multidimensional singular integral operators [38 (2)]. The investigation of systems of equations involving such operators has led to the notion of matrix symbol [59, 12 (14), 39, 41]. This notion plays an essential role not only for systems, but also for singular integral operators with piecewise-continuous (scalar) coefficients [44 (4)]. At the same time, attempts to introduce a (scalar or matrix) symbol for other algebras have failed. This is the case, for example, with algebras containing multidimensional Wiener-Hopf operators, one-dimensional singular integral operators whose coefficients have discontinuities of almost periodic type, or non-Carleman shift operators. It was conjectured that such algebras do not admit a symbol. In [44, 45] those operator algebras admitting a symbol were characterized and the conjecture was confirmed.

The main goal of this book is to discuss these and other related results.

In Chapter I a connection is established between the Fredholmness (invertibility, the value of the index) of operator matrices $[A_{mk}]_{m,k=1}^n$ and the Fredholmness (respectively, invertibility, the value of the index) of the determinant $det\ [A_{mk}]$.

In Chapter II one calculates the norm $\|S\|$ of the operator of singular integration S in the space L_p^n with power-type weight ρ, as well as the quotient norm of this operator, $|S| = \inf_T \|S + T\|$, where the infimum is taken over the set of all compact operators. In

---

* $[m(n)]$ denotes the nth entry in the list of references in $[m]$ (translator's note: for the reader's convenience we added to the bibliography some of the references labeled $[m(n)]$).

particular, if Γ_0 is the unit circle, $2 < p < \infty$, $t_0 \in \Gamma_0$, and $\rho(t) = |t - t_0|^\alpha$, then

$$\|S\|_{L_p^n(\Gamma_0,\rho)} = \|S\|_{L_p^n(\Gamma_0,\rho)} = \begin{cases} \cot\dfrac{\pi(p-1-\alpha)}{2p} & \text{if} \quad p-2 < \alpha < p-1 \quad , \\[2mm] \cot\dfrac{\pi}{2p} & \text{if} \quad 0 \le \alpha \le p-2 \quad , \\[2mm] \cot\dfrac{\pi(1+\alpha)}{2p} & \text{if} \quad -1 < \alpha < 0 \quad . \end{cases}$$

For the other values of α, the operator S in $L_p(\Gamma_0,\rho)$ is unbounded. For $1 < p < 2$

$$\|S\|_{L_p(\Gamma_0,\rho)} = \|S\|_{L_q^n(\Gamma_0,\rho^{1-q})} \ ,$$

where $p^{-1} + q^{-1} = 1$.

The results of the first two chapters are used in Chapter III, where a criterion for the Fredholmness of singular integral operators with piecewise-continuous matrix coefficients is established; the sufficient conditions for the Fredholmness of singular integral operators with L_∞ coefficients, obtained by I.B. Simonenko, are extended to the case of matrix (and operator) coefficients; and it is shown that these conditions are, in a certain sense, also necessary.

Conditions for the existence of a scalar symbol in algebras K of linear operators are investigated in Chapter IV. For symmetric algebras the condition is that the quotient-algebra K/T of K by the ideal T of compact operators be commutative. Symbols are constructed on algebras generated by singular integral operators with continuous coefficients on open contours, and also on algebras generated by Toeplitz matrices built from Fourier coefficients of piecewise-continuous functions.

It is shown that if the quotient algebra $K/\mathcal{R}(K)$ of the Banach algebra K by its radical $\mathcal{R}(K)$ is commutative, then K possesses a sufficient family of multiplicative functionals. The family $\{f_M\}_{M \in \alpha}$ of multiplicative functionals is said to be sufficient if an element $x \in K$ is invertible in K if and only if $f_M(x) \neq 0$ for all $M \in \alpha$.

In Chapter V one studies Banach algebras generated by one-dimensional singular operators with piecewise-continuous coefficients. It is proved that these algebras do not admit a scalar symbol. One constructs a matrix symbol and studies its properties. Singular integral operators with Carleman shift (and piecewise-continuous coefficients) are also considered.

Chapter VI discusses certain problems of the theory of Banach algebras connected with a matrix analog of the Gelfand transformation.

We say that the Banach algebra K possesses a sufficient family of n-dimensional representations if there exists a family $\{\nu_M\}_{M \in \alpha}$ of homomorphisms $\nu_M : K \to \mathbb{C}^{\ell \times \ell}$ (with $\ell = \ell(M) \le n$) such that

$$x \in GK \iff (\det \nu_M(x) \neq 0 \quad \forall M \in \alpha) \ ,$$

where GK designates the group of invertible elements of algebra K. It turns out that a Banach algebra K possesses a sufficient family of n-dimensional representations if and only if the quotient algebra $K/\mathcal{R}(K)$ of K by its radical $\mathcal{R}(K)$ is a PI-algebra (polynomial-identity algebra). The matrix analog of the Gelfand transformation considered here is constructed as follows.

Let K be a PI-algebra, $\mathcal{M}$ the set of all two-sided maximal ideals M of K, $K_M = K/M$, and a_M the element of K_M which contains a. Then

1) the quotient algebra K_M is isomorphic to $\mathbb{C}^{\ell \times \ell}$ ($\ell = \ell(M) \leq [d/2]$, where d designates the degree of the polynomial which gives an identity on K);

2) if $a \in K$, then
$$a \in GK \iff (a_M \in GK_M \quad \forall M \in \mathcal{M}) \;;$$

3) the radical of K coincides with the intersection of all maximal ideals of K.

In Chapter VII conditions are given for the existence of a matrix symbol in operator algebras, the order of the matrices (in a symbol) is sharpened, examples are presented of algebras with n-symbols as well as of algebras which admit no matrix symbol. In particular, a self-adjoint subalgebra of the algebra of bounded operators acting in a Hilbert space admits a symbol if and only if the quotient algebra $K/\mathcal{T}$ of K by the ideal of compact operators is a PI-algebra.

We conclude the book by formulating a number of open problems and briefly commenting on the literature. The list of references and the citations should serve the reader working on the text. The publications to which we refer are not necessarily first sources.

The author expresses his deep gratitude to A.S. Marcus, I.A. Fel'dman, I.E. Verbitsky and M.B. Abalovich for useful critical remarks made during the work on this monograph, and to N.S. Fedotova for helping with the preparation of the manuscript.

CONTENTS

CHAPTER I

MATRIX FREDHOLM OPERATORS

In this chapter a connection is established between the Fredholmness of the operator $A = [A_{mk}]_{m,k=1}^n$ in the Cartesian product $E^n = E \times \cdots \times E$ of n copies of a Banach space E and the Fredholmness of its determinant $det[A_{mk}]$ in E. Conditions are given under which the indices of these two operators coincide.

1. CONDITIONS FOR INVERTIBILITY OF MATRICES WITH ENTRIES IN A RING

Let K be an associative and, generally speaking, noncommutative ring with identity e, and let $K^{n \times n}$ designate the ring of matrices $[a_{mk}]_{m,k=1}^n$ with entries $a_{mk} \in K$, having the usual matrix operations of addition and multiplication. We let GK denote the group of invertible elements of K.

We shall examine the connection between the invertibility of a matrix $a = [a_{jk}]_{m,k=1}^n$ in the ring $K^{n \times n}$ and the invertibility of its determinant (more precisely, of one of its determinants) in K. If K is not commutative, then we must first establish a convention concerning the order in which the factors a_{mk} appear in the determinant $det[a_{mk}]$.

One can give simple examples in which the matrix a is invertible but none of its determinants is, and also examples in which all determinants are invertible, but the matrix a is not.

Example 1.1. Let $K = \mathbb{C}^{2 \times 2}$, $a_{11} = \begin{bmatrix} 0 & 0 \\ 0 & 1 \end{bmatrix}$, $a_{12} = \begin{bmatrix} 0 & 1 \\ 0 & 0 \end{bmatrix}$, $a_{21} = \begin{bmatrix} 0 & 0 \\ 1 & 0 \end{bmatrix}$,

$a_{22} = \begin{bmatrix} 1 & 0 \\ 0 & 0 \end{bmatrix}$, and $a = [a_{mk}] \in K^{2 \times 2}$. Here $a^2 = e$ and hence $a \in GK^{2 \times 2}$. But

$a_{11}a_{22} - a_{12}a_{21} = a_{22}a_{11} - a_{12}a_{21} = \begin{bmatrix} 1 & 0 \\ 0 & 0 \end{bmatrix} \notin GK$ and $a_{11}a_{22} - a_{21}a_{12} = a_{22}a_{11} -$

$a_{21}a_{12} = \begin{bmatrix} 0 & 0 \\ 0 & 1 \end{bmatrix} \notin GK$.

Example 1.2. Let $K = \mathbb{C}^{2 \times 2}$, $b_{11} = \begin{bmatrix} 2 & 0 \\ 0 & 1 \end{bmatrix}$, $b_{12} = \begin{bmatrix} 0 & 2 \\ 0 & 0 \end{bmatrix}$, $b_{21} = \begin{bmatrix} 0 & 0 \\ 2 & 0 \end{bmatrix}$,

$b_{22} = \begin{bmatrix} 1 & 0 \\ 0 & 2 \end{bmatrix}$, and $b = [b_{mk}] \in K^{2 \times 2}$. Here all four determinants of b are invertible in K, but $b \notin GK^{2 \times 2}$.

For $a, b \in K$ we denote $[a, b] = ab - ba$.

THEOREM 1.1. *Let $a_{mk} \in K$ $(m, k = 1, \ldots, n)$, and suppose that $[a_{mk}, a_{pq}] = 0$ for all $m, k, p, q = 1, \ldots, n$. Let $a = [a_{mk}] \in K^{n \times n}$ and $\Delta = det[a_{mk}]$. Then*

$$a \in GK^{n \times n} \Longleftrightarrow \Delta \in GK. \qquad (1.1)$$

PROOF. Let $\tilde{a}$ denote the adjoint of the matrix a: $\tilde{a} = [\tilde{a}_{mk}]$, where $\tilde{a}_{mk}$ is the cofactor of the entry a_{mk} in a. Using the commutativity assumption $[a_{mk}, a_{pq}] = 0$, one can readily check that $a\tilde{a} = [\delta_{mk}\Delta]_{m,k=1}^{n}$. It follows that if $\Delta \in GK$, then the matrix a is right invertible. The left invertibility of a is proved analogously.

Now let $a \in GK^{n \times n}$ and $a^{-1} = [c_{mk}]$. We show that the entries c_{mk} pairwise commute and also commute with the entries a_{pq} of the matrix a. Let r be any of the entries a_{pq} $(p, q = 1, \ldots, n)$. Then

$$[rc_{mk}] = [c_{mk}][a_{mk}][rc_{mk}] = [c_{mk}r][a_{mk}][c_{mk}] = [c_{mk}r], \qquad (1.2)$$

whence $[a_{pq}, c_{mk}] = 0$ for all p, q, m, k.

From this it follows that equality (1.2) holds also if r is any of the entries c_{pq} $(p, q = 1, \ldots, n)$, implying that $[c_{pq}, c_{mk}] = 0$ for all p, q, m, k. We have thus proved the asserted commutativity between the entries c_{mk}, and between the c_{mk} and the entries a_{pq}, in view of which the equalities $[c_{mk}][a_{mk}] = [a_{mk}][c_{mk}] = e$ imply that $det[a_{mk}]det[c_{mk}] = det[c_{mk}]det[a_{mk}] = e$. $\qquad \square$

Remark. For the validity of the equality $a\tilde{a} = [\delta_{mk}\Delta]$ $(\tilde{a}a = [\delta_{mk}\Delta])$ it suffices that the entries lying on distinct columns (respectively, rows) pairwise commute. This implies the following result:

THEOREM 1.2. *If the entries lying on distinct columns (rows) of the matrix* $a \in K^{n\times n}$ *pairwise commute and the element* $\Delta \in K$ *is right (respectively, left) invertible, then the matrix* a *is right (respectively, left) invertible.*

$\square$

The converse of this theorem is also valid:

THEOREM 1.3. *If the entries lying on distinct columns of the matrix* $a \in K^{n\times n}$ *pairwise commute and* a *is left invertible, then* $\det a$ *is left invertible.*

We first prove the following assertion:

LEMMA 1.1. *If the assumptions of Theorem 1.3 are in force, then for every minor* M_k *of order* k $(1 \le k < n)$ *of the matrix* a *one can find minors* M_{k+1}^p *of order* $k+1$ *of* a *and elements* $g_p \in K$ $(p = 1, \ldots, \ell)$ *such that*

$$\sum_{p=1}^{\ell} g_p M_{k+1}^p = M_p.$$

PROOF. We shall use the standard notation for the minors of a:

$$a\begin{pmatrix} i_1 & i_2 & \cdots & i_k \\ j_1 & j_2 & \cdots & j_k \end{pmatrix} = \det[a_{i_r j_s}]_{r,s=1}^k \qquad \left(\text{where } \begin{matrix} i_1 < i_2 < \cdots < i_k \\ j_1 < j_2 < \cdots < j_k \end{matrix}\right).$$

For simplicity we shall assume that

$$M_k = a\begin{pmatrix} 1 & 2 & \cdots & k \\ 1 & 2 & \cdots & k \end{pmatrix}$$

(this can always be achieved by permuting the rows and columns of a; under such a permutation the minors of a may only change sign).

Let

$$M_k^r = a\begin{pmatrix} 1 & \cdots & r-1 & r & \cdots & k \\ 1 & \cdots & r-1 & r+1 & \cdots & k+1 \end{pmatrix}, \qquad r = 1, \ldots, k,$$

and consider the column matrix $z = [z_m]_1^n \in K^{n+1}$, where

$$
\left.
\begin{aligned}
z_m &= (-1)^{k-m+1} M_k^m \ , \quad \text{if} \quad\quad m = 1, \ldots, k \ , \\
z_{k+1} &= M_k \ , \\
z_m &= 0 \ , \quad\quad\quad\quad\quad\quad \text{if} \quad\quad m > n+1 \ .
\end{aligned}
\right\} \tag{1.3}
$$

Let $y = az = [y_m]_1^n \ (\in K^{n+1})$. Then obviously

$$
y_i = \sum_{m=1}^{n} a_{im} z_m = \sum_{m=1}^{k} (-1)^{k-m+1} a_{im} M_k^m + a_{i,k+1} M_k \quad\quad i = 1, \ldots, n.
$$

From this we readily deduce that

$$
\left.
\begin{aligned}
y_i &= 0 \ , \quad\quad\quad\quad\quad\quad\quad\quad\quad\quad \text{if} \quad\quad i = 1, \ldots, k \ , \\
y_i &= a \begin{pmatrix} 1 & \cdots & k & i \\ 1 & \cdots & k & k+1 \end{pmatrix} , \quad \text{if} \quad\quad i > k \ .
\end{aligned}
\right\} \tag{1.4}
$$

Let the matrix $b = [b_{im}] \in K^{n \times n}$ be a left inverse of a. Then obviously $by = z$. In view of (1.3) and (1.4), this yields for the $(k+1)$-th entries of these column matrices the equalities

$$
\sum_{m=k+1}^{n} b_{k+1,m} a \begin{pmatrix} 1 & \cdots & k & m \\ 1 & \cdots & k & k+1 \end{pmatrix} = M_k.
$$

□

PROOF OF THEOREM 1.3.   We show that for every $k = 1, \ldots, n$ one can find minors M_k^i of order k of the matrix a and elements $g_k^i \in K$, $i = 1, \ldots, r_k$, such that

$$
\sum_{i=1}^{r_k} g_k^i M_k^i = e. \tag{1.5}
$$

Since

$$
\sum_{m=1}^{n} b_{1m} a_{m1} = e,
$$

where the matrix $[b_{im}]_{i,m=1}^n$ is a left inverse of a, equality (1.5) holds for $k = 1$. Applying Lemma 1.1 successively for $k = 1, \ldots, n-1$, we obtain (1.5) for $k = 2, \ldots, n$. For $k = n$ this equality states that $g_n^1 \det a = e$, i.e., $\det a$ is left-invertible as asserted.

□

Proceeding in a similar fashion one proves the following result:

THEOREM 1.4. *If the entries lying on distinct rows of the matrix $a \in K^{n \times n}$ pairwise commute and a is right invertible, then det a is right invertible.*

$\square$

We remark that if it is not assumed that the entries of the matrix pairwise commute then Theorem 1.1 fails even for triangular matrices. For such matrices the following result holds.

THEOREM 1.5. *Let $a = [a_{mk}] \in K^{n \times n}$ be a lower triangular matrix : $a_{mk} = 0$ if $m < k$. Then the following assertions are valid:*

1) *If $a_{mm} \in GK$, $m = 1, \ldots, n$, then $a \in GK^{n \times n}$;*

2) *If $a \in GK^{n \times n}$, then a_{11} (respectively, a_{nn}) is left (respectively, right) invertible.*

3) *If $a \in GK^{n \times n}$ and $a_{11}, \ldots, a_{pp} \in GK$ (where $p < n$), then $a_{p+1,p+1}$ is left invertible.*

4) *If $a \in GK^{n \times n}$ and $a_{pp}, \ldots, a_{nn} \in GK$ (where $p > 1$), then $a_{p-1,p-1}$ is right invertible.*

PROOF. 1) If $n = 2$, then

$$a^{-1} = \begin{bmatrix} a_{11}^{-1} & 0 \\ -a_{22}^{-1} a_{21} a_{11}^{-1} & a_{22}^{-1} \end{bmatrix}.$$

In the general case the proof proceeds by induction on n.

2) Let $c = [c_{mk}] = a^{-1}$. From the equality $ac = e$ (respectively, $ca = e$) it follows that $a_{11}c_{11} = e$ (respectively, $c_{nn}a_{nn} = e$).

3) In view of 1), the matrix $a_p = [a_{mk}]_{m,k=1}^{p}$ is invertible in $K^{p \times p}$. The matrix a can be represented in the form

$$a = \begin{bmatrix} a_p & 0 \\ c & b \end{bmatrix} = \begin{bmatrix} e_p & 0 \\ a_p^{-1} c & e_{n-p} \end{bmatrix} \begin{bmatrix} a_p & 0 \\ 0 & b \end{bmatrix},$$

where e_r denotes the identity element of the ring $K^{r \times r}$, $b = [a_{mk}]_{m,k=p+1}^n$, and $c = [a_{mk}]$ $(p+1 \le m \le n, 1 \le k \le p)$. Since $a \in GK^{n \times n}$, it follows that $b \in GK^{(n-p) \times (n-p)}$ and hence, in view of 2), that $a_{p+1,p+1} \in GK$.

Assertion 4) is proved similarly.

$\square$

As we saw above (Examples 1.1 and 1.2), in the general case there is no direct connection between the invertibility of a matrix $[a_{mk}] \in K^{2 \times 2}$ and the invertibility of the element $\Delta = a_{11}a_{22} - a_{21}a_{12} \in K$. It turns out, however, that the invertibility of Δ is a necessary and sufficient condition for the invertibility of the 3×3 matrix

$$a = \begin{bmatrix} e & 0 & -a_{22} \\ 0 & e & a_{12} \\ a_{11} & a_{21} & 0 \end{bmatrix}.$$

This is a consequence of the following theorem:

THEOREM 1.6. *Let $a_0, b_m, c_m \in K$, $m = 1, \ldots, n$, and put*

$$a = \begin{bmatrix} e & 0 & \cdots & 0 & c_1 \\ 0 & e & \cdots & 0 & c_2 \\ \cdot & \cdot & \cdots & \cdot & \cdot \\ 0 & 0 & \cdots & e & c_n \\ b_1 & b_2 & \cdots & b_n & a_0 \end{bmatrix} \in K^{n+1 \times n+1}.$$

Then $a \in GK^{(n+1) \times (n+1)}$ if and only if $a_0 - \sum_{m=1}^n b_m c_m \in GK$.

We note that $a_0 - \sum_{m=1}^n b_m c_m = \det a$ (more precisely, one of the determinants of a).

PROOF. It is readily verified that the following equality holds:

$$\begin{bmatrix} e & 0 & \cdots & 0 & c_1 \\ 0 & e & \cdots & 0 & c_2 \\ \cdot & \cdot & \cdots & \cdot & \cdot \\ 0 & 0 & \cdots & e & c_n \\ b_1 & b_2 & \cdots & b_n & a_0 \end{bmatrix}$$

$$= \begin{bmatrix} e & 0 & \cdots & 0 & 0 \\ 0 & e & \cdots & 0 & 0 \\ \cdot & \cdot & \cdots & \cdot & \cdot \\ 0 & 0 & \cdots & e & 0 \\ b_1 & b_2 & \cdots & b_n & e \end{bmatrix} \begin{bmatrix} e & 0 & \cdots & 0 & 0 \\ 0 & e & \cdots & 0 & 0 \\ \cdot & \cdot & \cdots & \cdot & \cdot \\ 0 & 0 & \cdots & e & 0 \\ 0 & 0 & \cdots & 0 & \Delta \end{bmatrix} \begin{bmatrix} e & 0 & \cdots & 0 & c_1 \\ 0 & e & \cdots & 0 & c_2 \\ \cdot & \cdot & \cdots & \cdot & \cdot \\ 0 & 0 & \cdots & e & c_n \\ 0 & 0 & \cdots & 0 & e \end{bmatrix}, \quad (1.6)$$

where $\Delta = a_0 - \sum b_m c_m$. Since the factors in the right-hand side of (1.6) are invertible in $K^{(n+1)\times(n+1)}$, we see that $a \in GK^{(n+1)\times(n+1)}$ if and only if $\Delta \in GK$.

$\square$

To conclude this section we consider the following problem: Let $x_{m\ell}$, $m = 1,\ldots,k_j$; $\ell = 1,\ldots,r$, be elements of the ring K and

$$\Delta = \sum_{n=1}^{k} x_{m1} x_{m2} \cdots x_{mr}. \tag{1.7}$$

Is there a matrix $a = [a_{pq}] \in K^{s\times s}$ whose entries belong the the set $\{0, e, x_{11}, \ldots, x_{kr}\}$, such that $a \in GK^{s\times s}$ if and only if $\Delta \in GK$?

We show that the answer is affirmative. Consider the matrices

$$z = \begin{bmatrix} e_k & b_1 & 0 & \cdots & 0 & 0 \\ 0 & e_k & b_2 & \cdots & 0 & 0 \\ \cdot & \cdot & \cdot & \cdots & \cdot & \cdot \\ 0 & 0 & 0 & \cdots & e_k & b_r \\ 0 & 0 & 0 & \cdots & 0 & e_k \end{bmatrix} \in K^{(r+1)k\times(r+1)k},$$

where e_m is the identity of the ring $K^{m\times m}$,

$$b_j = \begin{bmatrix} -x_{1j} & 0 & 0 & \cdots & 0 \\ 0 & -x_{2j} & 0 & \cdots & 0 \\ \cdot & & \cdot & \cdots & \cdot \\ 0 & 0 & 0 & \cdots & -x_{kj} \end{bmatrix},$$

and also the column matrix

$$x = \begin{bmatrix} 0 \\ \vdots \\ 0 \\ -e \\ \vdots \\ -e \end{bmatrix} \begin{array}{l} \left.\vphantom{\begin{matrix}0\\ \vdots \\ 0\end{matrix}}\right\} kr \\ \left.\vphantom{\begin{matrix}-e\\ \vdots \\ -e\end{matrix}}\right\} k \end{array}$$

and the two row matrices

$$y = \begin{bmatrix} \overbrace{e\ e\ \ldots\ e}^{k} & \overbrace{0\ \ldots\ 0}^{kr} \end{bmatrix}$$

and

$$u = [M_0 \quad M_1 \quad \ldots \quad M_r]$$

where $M_0 = [e \quad e \quad \ldots \quad e]$ and for each $j = 1, 2, \ldots, r$

$$M_j = [x_{11} \cdot x_{12} \cdot \ldots \cdot x_{1j}, x_{21} \cdot x_{22} \cdot \ldots \cdot x_{2j}, \ldots, x_{k1} \cdot x_{k2} \cdot \ldots \cdot x_{kj}].$$

THEOREM 1.7. *Let* $x_{j\ell} \in K$ $(j = 1, \ldots, k; \ \ell = 1, \ldots, r)$ *and* $\Delta = \sum_{j=1}^{k} x_{j1} x_{j2} \ldots x_{jr}$. *Then:*

$$\begin{bmatrix} z & x \\ y & 0 \end{bmatrix} = \begin{bmatrix} e_{k(r+1)} & 0 \\ u & e \end{bmatrix} \begin{bmatrix} e_{k(r+1)} & 0 \\ 0 & \Delta \end{bmatrix} \begin{bmatrix} z & x \\ 0 & e \end{bmatrix} \tag{1.8}$$

PROOF. It is readily verified that $uz = y$ and $ux + \Delta = 0$, which is equivalent to (1.8).

□

We remark that the outer factors in the right-hand side of equality (1.8) are triangular matrices with identity elements on the diagonal and hence invertible in the ring $K^{s \times s}$ (where $s = rk + k + 1$).

We call the matrix in the right-hand side of (1.8) the *linear dilation* of the element Δ corresponding to representation (1.7), and denote it by $\Xi(\Delta)$.

Theorem 1.7 has the following corollary:

COROLLARY 1.1. *In order that the element* Δ *defined by equality* (1.7) *be left (right) invertible in* K, *it is necessary and sufficient that its linear dilation* $\Xi(\Delta)$ *be left (respectively, right) invertible in* $K^{s \times s}$.

2. PROPERTIES OF MATRIX Φ-OPERATORS

Let $\mathcal{L}(E)$ denote the algebra of all bounded linear operators acting in the Banach space E and $\mathcal{T} = \mathcal{T}(E)$ the closed two-sided ideal of all compact operators.

We remind the reader that an operator $A \in \mathcal{L}(E)$ is called a Φ-*operator* or a *Fredholm operator* if $\overline{Im\ A} = Im\ A$ (i.e., A is normally solvable), $dim\ Ker\ A < \infty$, and $dim\ Ker\ A^* < \infty$. If $\overline{Im\ A} = Im\ A$ and $dim\ Ker\ A < \infty$ $(dim\ Ker\ A^* < \infty)$, then A is called a Φ_+- (respectively, Φ_--) operator. We denote the set of all Φ- (Φ_+-, Φ_--) operators in E by $\Phi(E)$ (respectively, $\Phi_+(E)$, $\Phi_-(E)$). An operator $A \in \mathcal{L}(E)$ belongs to $\Phi(E)$ if and only if it is regularizable, i.e., if and only if there is an operator $R \in \mathcal{L}(E)$ such that $AR - I \in \mathcal{T}$ and $RA - I \in \mathcal{T}$.

The *index* of the Φ-operator A is, by definition, the number $Ind\ A = dim\ Ker\ A - dim\ Ker\ A^*$.

Let $E^n = \underbrace{E \times E \times \ldots \times E}_{n}$. It is readily verified that $\mathcal{L}(E^n) = \mathcal{L}(E)^{n \times n}$ and $\mathcal{T}(E^n) = \mathcal{T}(E)^{n \times n}$. We represent each operator $A \in \mathcal{L}(E^n)$ in the (block) form $A = [A_{jk}]_{j,k=1}^n$, where $A_{jk} \in \mathcal{L}(E)$ $(j, k = 1, \ldots, n)$. In this section we prove, for certain classes of operators A_{jk}, that A is a Φ- (Φ_+-, Φ_--) operator in E^n if and only if $det[A_{jk}]_{j,k=1}^n$ is a Φ- (respectively, Φ_+-, Φ_--) operator in E.

THEOREM 2.1.   *Let* $A_{jk} \in \mathcal{L}(E)$, $(j, k = 1, \ldots, n)$, $[A_{jk}, A_{pq}] \in \mathcal{T}(E)$, $j, k, p, q = 1, \ldots, n$, $A = [A_{jk}]_{j,k=1}^n$, *and* $\Delta = det[A_{jk}]$. *Then*

1. $A \in \Phi(E^n) \Longleftrightarrow \Delta \in \Phi(E)$;

2. $A \in \Phi_\pm(E^n) \Longleftrightarrow \Delta \in \Phi_\pm(E)$;

3.   *A is left (right) regularizable if and only if* Δ *is left (respectively, right) regularizable.*

Remark.   When one writes the determinant $det[A_{jk}]$ the order of the factor is irrelevant since the possible determinants differ from one another by compact terms.

PROOF.   Let $\tilde{A}$ be the adjoint of the matrix $A = [A_{jk}]$. Then $A\tilde{A} = \tilde{A}A = [\delta_{jk}\Delta]$. If $\Delta \in \Phi_+(E)$ $(\Delta \in \Phi_-(E))$, then $\tilde{A}A \in \Phi_+(E^n)$ (respectively, $\tilde{A}A \in \Phi_-(E^n)$). It follows from known properties of $\Phi_\pm$-operators that $A \in \Phi_+(E^n)$ (respectively, $A \in \Phi_-(E^n)$) (see [50(9)]). Applying Theorems 1.1–1.4 to the quotient algebra $\mathcal{L}(E)/\mathcal{T}(E)$ one can readily verify assertions 1) and 3) of the theorem. It remains to show that $A \in$

$\Phi_\pm(E^n) \implies \Delta \in \Phi_\pm(E)$. To this end we need two lemmas.

LEMMA 2.1. *If $A \in \Phi_+(E)$, then there exists an operator $T_0 \in T(E)$ and a number $\delta > 0$ such that*

$$\|Ax\| + \|T_0 x\| \geq \delta \|x\| . \tag{2.1}$$

Conversely, if for $A \in \mathcal{L}(E)$ there are operators $T_m \in T(E)$ $(m = 1,\ldots,k)$ and a $\delta > 0$ such that

$$\|Ax\| + \sum_{m=1}^{k} \|T_m x\| \geq \delta \|x\| , \tag{2.2}$$

then $A \in \Phi_+(E)$.

PROOF OF LEMMA 2.1. If $A \in \Phi_+(E)$, then, as is readily verified, inequality (2.1) holds for any projector T_0 onto the subspace $Ker\, A$.

Now suppose that inequality (2.2) is satisfied, but $A \notin \Phi_+(E)$. Then there is an infinite-dimensional subspace X of E such that $2\|Ax\| \leq \delta\|x\|$ for all $x \in X$ (see, e.g., [31, Theorems IV.5.9. and IV.5.10]), and hence

$$\sum_{m=1}^{k} \|T_m x\| \geq \frac{\delta}{2} \|x\| . \tag{2.3}$$

Let $\{x_n\}_1^\infty$ be an arbitrary bounded sequence of elements of X. Then it contains a subsequence $\{x_{nj}\}$ such that $\{T_m x_{nj}\}_{j=1}^\infty$ $(m = 1,\ldots,k)$ are Cauchy sequences. But then $\{x_{nj}\}$ is itself a Cauchy sequence by (2.3), which contradicts the fact that X is infinite-dimensional.

$\square$

For the next lemma let us agree, for the sake of definiteness, to form determinants according to the rule

$$det[A_{mk}]_{m,k=1}^n = \sum_\sigma (-1)^\sigma A_{n\sigma(n)} \cdots A_{1\sigma(1)} , \tag{2.4}$$

i.e., to write the factors (in each term) in the order of row numbers.

LEMMA 2.2.   *Let $A \in \Phi_+(E^n)$ and $[A_{mk}, A_{pq}] \in \mathcal{T}(E)$ for $k \neq q$. Then for every minor M_k of order k of the matrix A there are minors M_{k+1}^p $(p = 1, \ldots, s)$ of order $k + 1$ of A, operators $T_m \in \mathcal{T}(E)$ $(m = 1, \ldots, t)$, and a $\delta > 0$ such that*

$$\sum_{p=1}^{s} \|M_{k+1}^p x\| + \sum_{m=1}^{t} \|T_m x\| \geq \delta \|M_k x\| \quad (x \in E) \ .$$

PROOF. For simplicity we shall assume that

$$M_k = A \begin{pmatrix} 1 \ldots k \\ 1 \ldots k \end{pmatrix} \ .$$

We put

$$\left. \begin{aligned} M_k^r &= A \begin{pmatrix} 1 \ldots r - 1 & r & \ldots & k \\ 1 \ldots r - 1 & r+1 & \ldots & k+1 \end{pmatrix} , \\[2mm] z_j &= (-1)^{k-j+1} M_k^j x \ , \quad j = 1, \ldots, k \ , \\[2mm] z_{k+1} &= M_k x \ , \\[2mm] &\text{and} \\[2mm] z_j &= 0 \quad (j = k + 2, \ldots, n) \ , \end{aligned} \right\} \qquad (2.5)$$

and consider the vector $Az = (y_j)_{j=1}^n \in E^n$, where $z = (z_j)_{j=1}^n$. Obviously,

$$y_j = \sum_{m=1}^{n} A_{jm} z_m = \sum_{m=1}^{k} (-1)^{k-m+1} A_{jm} M_k^m x + A_{j,k+1} M_k x \ , \quad (j = 1, \ldots, n) \ .$$

Taking into account the assumption that $[A_{mk}, A_{pq}] \in \mathcal{T}(E)$ for $k \neq q$ and definition (2.4), we get

$$\left. \begin{aligned} y_j &= S_j x \ , \quad j = 1, \ldots, k \ , \\[2mm] y_j &= M_{k+1}^j \ , \quad j = k + 1, \ldots, n \ . \end{aligned} \right\} \qquad (2.6)$$

Here $S_j \in \mathcal{T}(E)$ and

$$M_{k+1}^j = A \begin{pmatrix} 1 \ldots k & j \\ 1 \ldots k & k+1 \end{pmatrix} .$$

By Lemma 2.1, there exists an operator $T_0 \in \mathcal{T}(E^n)$ and a $\delta_0 > 0$ such that

$$\|Au\| + \|T_0 u\| \geq \delta_0 \|u\| \quad (u \in E^n) . \tag{2.7}$$

If we now substitute z for u we get, in view of (2.5) and (2.6),

$$\sum_{j=k+1}^{n} \|M_{k+1}^j x\| + \sum_{j=1}^{k} \|S_j x\| + \sum_{j=1}^{n} \|R_j x\| \geq \delta_0 \|z\| \geq \delta_0 \|z\| \geq \delta_0 \|M_k\|$$

with $R_j \in \mathcal{T}(E)$. This proves the lemma.

We now return to the proof of Theorem 2.1. Letting $u = (x, 0, \ldots, 0)$ in (2.7), we have

$$\sum_{m=1}^{n} \|A_{m1} x\| + \sum_{m=1}^{n} \|T_{m1} x\| \geq \delta_0 \|x\| ,$$

where $T_{m1} \in \mathcal{T}(E)$. Applying Lemma 2.2 successively for $k = 1, 2, \ldots, n-1$, we finally obtain

$$\|\Delta x\| + \sum_{m=1}^{m} \|T_m x\| \geq \delta \|x\| \quad (x \in E) ,$$

where $T_m \in \mathcal{T}(E)$ and $\delta > 0$. By Lemma 2.1 this means that $\Delta \in \Phi_+(E)$.

The assertion concerning Φ-operators is deduced from the one just proved by passing to conjugate operators.

$\square$

THEOREM 2.2. Let $A_{mk} \in \mathcal{L}(E)$ $(m, k = 1, \ldots, n)$ and suppose that $A_{mk} \in \mathcal{T}(E)$ for $m < k$. Let $A = [A_{mk}]_{m,k=1}^{n}$. If:

1) $A_{mm} \in \Phi(E)$, then $A \in \Phi(E^n)$;

2) $A \in \Phi(E^n)$, then $A_{11} \in \Phi_-(E)$ and $A_{nn} \in \Phi_+(E)$;

3) $A \in \Phi(E^n)$ and $A_{11}, \ldots, A_{pp} \in \Phi(E)$ (with $p < n$)), then $A_{p+1,p+1} \in \Phi_-(E)$;

4) $A \in \Phi(E^n)$ and $A_{pp}, \ldots, A_{nn} \in \Phi(E)$ (with $p > 1$), then $A_{p-1,p-1} \in \Phi_+(E)$.

PROOF. It suffices to apply Theorem 1.5 to the quotient algebra $\mathcal{L}(E)/\mathcal{T}(E)$. Proceeding in this manner we actually obtain more information on the operators A_{11}, A_{nn}, A_{p+1}, A_{p+1} and $A_{p-1,p-1}$ than in the assertions of the theorem, namely, they admit left or right regularizers, depending on the case.

□

THEOREM 2.3. Let $B_m, C_m, A_0 \in \mathcal{L}(E)$, $\Delta = A_0 - \Sigma_{m=1}^{n} B_m C_m$, and

$$
A = \begin{bmatrix}
I & 0 & \ldots & 0 & C_1 \\
0 & I & \ldots & 0 & C_2 \\
\cdot & \cdot & \ldots & \cdot & \cdot \\
0 & 0 & \ldots & I & C_n \\
B_1 & B_2 & \ldots & B_n & A_0
\end{bmatrix} .
$$

Then the operator A is normally solvable in E^{n+1} if and only if the operator Δ is normally solvable in E. Moreover, dim Ker A = dim Ker Δ and dim Coker A = dim Coker Δ.

PROOF. It suffices to use equality (1.6) and the obvious fact that the normal solvability of a bounded linear operator A is preserved under its multiplication by invertible operators and so are the numbers dim Ker A and dim Coker A.

□

COROLLARY 2.1. Under the assumptions of Theorem 2.3

$$
A \in \Phi_+(E^{n+1}) \Longleftrightarrow \Delta \in \Phi_+(E) \quad and \quad A \in \Phi_-(E^{n+1}) \Longleftrightarrow \Delta \in \Phi_-(E) .
$$

□

THEOREM 2.4. Let $A_{mn} \in \mathcal{L}(E)$, $m = 1, \ldots, k$; $n = 1, \ldots, r$, and

$$
A = \sum_{m=1}^{k} A_{m1} A_{m2} \ldots A_{mr} .
$$

Then:

1) *A is normally solvable in E if and only if $\Xi(A)$ is normally solvable in E^s, where $s = kr + k + 1$;*

2) *dim Ker A = dim Ker $\Xi(A)$ and dim Coker A = dim Coker $\Xi(A)$;*

3) *$A \in \Phi_+(E) \Longleftrightarrow \Xi(A) \in \Phi_+(E^s)$ and $A \in \Phi_-(E) \Longleftrightarrow \Xi(A) \in \Phi_-(E^s)$.*

PROOF. It suffices to use equality (1.8).

$\square$

3. THE INDEX OF MATRIX Φ-OPERATORS

In this section we discuss the connection between the index of the operator $A = [A_{mk}]_{m,k=1}^n$, acting in E^n, and the index of its determinant. In Sec.2 we showed that if $[A_{mk}, A_{pq}] \in \mathcal{T}(E)$ for $m, k, p, q = 1, \ldots, n$, then $A \in \Phi(E^n)$ if and only if $\Delta = det[A_{mk}] \in \Phi(E)$. In this case it is natural to ask whether the indices of the Φ-operators A and Δ coincide. Generally speaking, the answer is negative. As an example, we consider multidimensional singular integral operators

$$(A_{mk}\varphi)(x) = a_{mk}(x)\varphi(x) + \int_{\mathbb{R}^d} \frac{f_{mk}(x,y)\varphi(y)dy}{|y-x|^d} \qquad (3.1)$$

with continuous symbols in the space $E = L_p(\mathbb{R}^d)$ (see [59]). It was shown in [79] that for $n > 1$ the index of the operator $A = [A_{mk}]_{m,k=1}^n$ may be different from zero. The A_{mk} commute modulo compact operators. The determinant Δ can be represented in the form $\Delta = A_0 + T$, where A is an operator of the form (3.1) and $T \in \mathcal{T}(E)$. In [59] it is proved that $Ind\ det\ A_0 = 0$. We can thus construct an example of an operator $A = [A_{mk}] \in \Phi(E^n)$ such that $[A_{mk}, A_{pq}] \in \mathcal{T}(E)$, but $Ind\ A \neq Ind\ det\ A$.

We give a number of conditions sufficient for the coincidence of $Ind\ A$ and $Ind\ det\ A$.

Let $\mathcal{K}$ be a subalgebra of $\mathcal{L}(E)$ with the following properties:

$\mathcal{T}(E) \in \mathcal{K}$; $[A, B] \in \mathcal{T}(E)$ for all $A, B \in \mathcal{K}$;   the set $\Phi(E) \cap \mathcal{K}$ is dense in $\mathcal{K}$.

THEOREM 3.1. *Let* $A_{mk} \in K$, $m, k = 1, \ldots, n$. *If* $A = [A_{mk}] \in \Phi(E^n)$, *then*
Ind $A = Ind \ det[A_{mk}]$.

PROOF. We proceed by induction on n. Suppose that $A \in \Phi(E^n)$. Then it
follows from the properties of algebra K that there is an operator $A_0 \in \Phi(E) \cap K$ such
that the norm $\|A_0 - A_{11}\|$ is small enough, in particular, so small that the indices of the
operators A and $\Delta = det[A_{mk}]$ are not affected on replacing A_{11} by A_0. Thus, we may
assume that $A_{11} \in \Phi(E) \cap K$. Let R be a regularizer for A_{11}. We let $\tilde{R}$ denote the diagonal
matrix $\tilde{R} = diag(R, I, \ldots, I)$. Then

$$
\tilde{R}A = \begin{bmatrix} RA_{11} & RA_{12} & \ldots & RA_{1n} \\ A_{21} & A_{22} & \ldots & A_{2n} \\ \cdot & \cdot & \ldots & \cdot \\ A_{n1} & A_{n2} & \ldots & A_{nn} \end{bmatrix}
$$

$$
\doteq \begin{bmatrix} I & 0 & \ldots & 0 \\ A_{21} & I & \ldots & 0 \\ \cdot & \cdot & \ldots & \cdot \\ A_{n1} & 0 & \ldots & I \end{bmatrix} \begin{bmatrix} I & 0 & \ldots & 0 \\ 0 & B_{22} & \ldots & B_{2n} \\ \cdot & \cdot & \ldots & \cdot \\ 0 & B_{n2} & \ldots & B_{nn} \end{bmatrix} \begin{bmatrix} I & RA_{12} & \ldots & RA_{1n} \\ 0 & I & \ldots & 0 \\ \cdot & \cdot & \ldots & \cdot \\ 0 & 0 & \ldots & I \end{bmatrix} + T
$$

$$
= S_1 S_2 S_3 + T \, ,
$$

(3.2)

where $B_{k\ell} = A_{k\ell} - A_{k1} R A_{1\ell}$, $k, \ell = 2, \ldots, n$, and $T \in \mathcal{T}(E^n)$. Notice that

$$
A_{kj} R - R A_{kj} = R A_{11} A_{kj} R - R A_{kj} + T_{kj} = R A_{kj} A_{11} R - R A_{kj} + T'_{kj} = T''_{kj}
$$

with $T_{kj}, T'_{kj}, T''_{kj} \in \mathcal{T}(E)$. It follows that the matrix $B = [B_{k\ell}]^n_2$ can be represented in
the form

$$
B = [\delta_{k\ell} R][A_{11} A_{k\ell} - A_{k1} A_{1\ell}] + T_0
$$

with $T_0 \in \mathcal{T}(E^{n-1})$. By the inductive hypothesis

$$
Ind[A_{11} A_{k\ell} - A_{k\ell} A_{1\ell}]^n_2 = Ind \ det[A_{11} A_{k\ell} - A_{k\ell} A_{1\ell}]^n_2 \, .
$$

It is readily checked that

$$Ind[\delta_{k\ell}R]_2^n = (n-1)Ind\ R = Ind\ det[\delta_{k\ell}R]_2^n\ .$$

Therefore, $Ind\ B = Ind\ det\ B$.

The operators S_1 and S_3 are invertible, and hence

$$Ind\ \tilde{R}A = Ind\ S_2 = Ind\ B = Ind\ det[B_{k\ell}]_2^n = Ind\ det\ S_2\ .$$

But $det\ \tilde{R}\ det\ A = det\ S_2 + T_0$ (see the remark just below Theorem 2.1), where $T_0 \in \mathcal{T}(E)$ and $Ind\ \tilde{R} = Ind\ det\ \tilde{R}$. It follows that

$$Ind\ A = Ind\ S_2\ -\ Ind\ \tilde{R} = Ind\ det\ S_2 - Ind\ det\ \tilde{R}$$

$$= Ind\ det\ \tilde{R} + Ind\ det\ A - Ind\ det\ \tilde{R} = Ind\ det\ A\ .$$

$\square$

In the next two theorems it is not required that the entries of the matrices A commute modulo compact operators.

Let E_1 and E_2 be Banach spaces, $E = E_1 \times E_2$, $A_{mk} \in \mathcal{L}(E_k, E_m)$, $m, k = 1, 2$, and $A = [A_{mk}]_{m,k=1}^2$.

THEOREM 3.2. *Suppose that $A_{11} \in \Phi(E_1)$ and let R be a regularizer for A_{11}. Then*

$$A \in \Phi(E) \Longleftrightarrow A_{22} - A_{21}RA_{12} \in \Phi(E)\ .$$

Moreover, if $A \in \Phi(E)$, then

$$Ind\ A = Ind[A_{11}(A_{22} - A_{21}RA_{12})]\ .$$

PROOF. The assertions of the theorem follow from the readily verifiable equality

$$A = \begin{bmatrix} I & 0 \\ A_{21}R & I \end{bmatrix} \begin{bmatrix} A_{11} & 0 \\ 0 & A_{22} - A_{21}RA_{12} \end{bmatrix} \begin{bmatrix} I & RA_{12} \\ 0 & I \end{bmatrix} + T \qquad (3.3)$$

with $T \in \mathcal{T}$, and the invertibility of the two extreme factors.

COROLLARY 3.1. *Suppose that* $A = [A_{mk}] \in \mathcal{L}(E^2)$, $A_{11} \in \Phi(E)$, *and at least one of the operators* $[A_{11}, A_{12}]$, $[A_{11}, A_{21}]$ *is compact. Then*

$$A \in \Phi(E^2) \Longleftrightarrow A_{11}A_{22} - A_{21}A_{12} \in \Phi(E) .$$

Moreover, if $A \in \Phi(E^2)$, *then* $Ind\ A = Ind(A_{11}A_{22} - A_{21}A_{12})$.

□

We remark that in this corollary the determinant $A_{11}A_{22} - A_{21}A_{12}$ cannot be replaced by $A_{11}A_{22} - A_{12}A_{21}$ or $A_{22}A_{11} - A_{12}A_{21}$. To see this, take $E = \ell_2$, $A_{11} = I$, $A_{21}\{\xi_n\} = \{\xi_{2n}\}$, $A_{12}\{\xi_n\} = \{0, \xi_1, 0, \xi_2, 0, \xi_3, \ldots\}$, $A_{22} = 0$, and $A = [A_{mk}]^2_{m,k=1}$. In this example $A_{11}A_{22} - A_{21}A_{12} = I$, but $A_{mm}A_{kk} - A_{12}A_{21} \notin \Phi(E)$ if $m \neq k$.

THEOREM 3.3. *Let* $A = [A_{mk}]$ *be a triangular matrix. If* $A_{mm} \in \Phi(E)$ *for every* m, *then* $Ind\ A = Ind\ det\ A$.

PROOF. We can represent the operator A in the form $A = BC + T$, where $C = diag(A_{11}, \ldots, A_{nn})$, B is a triangular matrix whose diagonal entries are all equal to the identity, and $T \in \mathcal{T}(E^n)$. Hence, $Ind\ A = \Sigma\ Ind\ A_{mm} = Ind\ det\ A$. □

THEOREM 3.4. *Let* E *be a Hilbert space,* $\mathcal{T}_1 \subset \mathcal{T}(E)$ *the set of trace-class operators (i.e., operators admitting a representation* $Ax = \Sigma_{n=1}^{\infty} f_n(x) y_n$ *with* $y_n \in E$, $f_n \in E^*$, *and* $\Sigma_n \|f_n\|\|y_n\| < \infty$), $A_{mk} \in \mathcal{L}(E)$, *and* $A = [A_{mk}] \in \Phi(E^n)$. *If* $[A_{mk}, A_{pq}] \in \mathcal{T}_1$ *for all* $m, k, p, q = 1, \ldots, n$, *then* $Ind\ A = Ind\ det\ A$.

PROOF. Let $\Delta = det\ A$. By Theorem 2.1, $\Delta \in \Phi(E)$, and hence it admits a regularizer $M \in \mathcal{T}_1$. Set $D = diag(M, I, \ldots, I)$ and $DA = B = [B_{mk}]$. It is readily checked that $[B_{mk}, B_{pq}] \in \mathcal{T}_1$. Since $Ind\ B = Ind\ D + Ind\ A = Ind\ M + Ind\ A = Ind\ A - Ind\ det\ A$, it remains to show that $Ind\ B = 0$.

To this end we shall need the cofactors of the entries of the matrix B. Let us agree here to write the factors B_{mk} in the increasing order of the first index, so that the cofactor R_{mk} of the entry B_{mk} will be an algebraic sum of terms of the form

$B_{1\sigma(1)} \cdots B_{m-1\sigma(m-1)} B_{m+1\sigma(m+1)} \cdots B_{n\sigma(n)}$, where $\sigma(1), \ldots, \sigma(m-1), \sigma(m+1), \ldots, \sigma(n)$
is some permutation of the numbers $1, \ldots, k-1, k+1, \ldots, n$. We let R denote the matrix
adjoint of the operator matrix B, i.e., $R = [R_{mk}]_{m,k=1}^n$. Since $RB - I \in T_1$ and $BR - I \in T_1$,
we have, by a formula of B.V. Fedosov (see [50(9)]), $Ind\ B = tr(BR - RB)$, where $tr\ X$
designates the trace of the operator $X \in T_1$. Moreover, since $tr[C_{mk}] = tr\ C_{11} + \ldots + tr\ C_{nn}$
for every matrix C,

$$Ind\ B = tr \sum_{m,k=1}^n (B_{mk} R_{mk} - R_{km} B_{km}) \ .$$

We can reexpress this equality as

$$Ind\ B = \sum_\sigma (-1)^{p(\sigma)} tr \left(\sum_{k=2}^{n-1} B_{k\sigma(k)} B_{1\sigma(1)} \cdots B_{k-1\sigma(k-1)} B_{k+1\sigma(k+1)} \cdots B_{n\sigma(n)} \right.$$

$$+ B_{n\sigma(n)} B_{1\sigma(1)} \cdots B_{n-1\sigma(n-1)} - B_{2\sigma(2)} \cdots B_{n\sigma(n)} B_{1\sigma(1)}$$

$$\left. - \sum_{k=2}^{n-1} B_{1\sigma(1)} \cdots B_{k-1\sigma(k-1)} B_{k+1\sigma(k+1)} \cdots B_{n\sigma(n)} B_{k\sigma(k)} \right) \ .$$

Here the sum $\sum_\sigma$ is taken over all permutations σ of the numbers $1, \ldots, n$, and $p(\sigma)$ denotes
the parity of σ. Hence, to prove that $Ind\ B = 0$ it suffices to establish the equality

$$tr \left(\sum_{k=2}^{n-1} S_k S_1 \ldots S_{k-1} S_{k+1} \ldots S_n + S_n S_1 \ldots S_{n-1} \right.$$

$$\left. - \sum_{k=2}^{n-1} S_1 \ldots S_{k-1} S_{k+1} \ldots S_n S_k - S_2 \ldots S_n S_1 \right) = 0 \tag{3.4}$$

for arbitrary operators $S_j \in \mathcal{L}(E)$ such that $[S_m, S_k] \in T_1$ for $m, k = 1, \ldots, n$. Since
$tr(AB) = tr(BA)$ for all $A \in T_1$ and $B \in \mathcal{L}(E)$, we have

$$tr(S_k S_1 \ldots S_{k-1} S_{k+1} \ldots S_n - S_k S_{k+1} \ldots S_n S_1 \ldots S_k)$$

$$= tr(S_1 \ldots S_{k-1} S_{k+1} \ldots S_n S_k - S_{k+1} \ldots S_n S_1 \ldots S_k) \ .$$

Adding these equalities for $k = 2, \ldots, n - 1$ we obtain (3.4). $\square$

The proof of Theorem 3.4 permits us to generalize it to the case of Banach spaces as follows:

THEOREM 3.5.   *Let I be a two-sided ideal in the algebra $\mathcal{L}(E)$ and suppose that on I there is defined a functional $tr(A)$ which coincides with the trace on finite-rank operators A, and such that $tr(AB) = tr(BA)$ for all $A \in I$ and $B \in \mathcal{L}(E)$. If $A = [A_{mk}] \in \Phi(E^n)$ and $[A_{mk}, A_{pq}] \in I$ for $m, k, p, q = 1, \ldots, n$, then Ind $A =$ Ind det A.*

$\square$

In particular, one can take for I the ideal of all finite-rank operators.

CHAPTER II

EXACT CONSTANTS IN BOUNDEDNESS THEOREMS
FOR SINGULAR INTEGRAL OPERATORS

Let Γ be a rectifiable curve in the complex plane $\mathbb{C}$ and let S_Γ (or simply S) denote the operator of singular integration along Γ:

$$(S_\Gamma\varphi)(t) = \frac{1}{\pi i}\int_\Gamma \frac{\varphi(\tau)d\tau}{\tau - t} \qquad (t \in \Gamma) \ .$$

Here the integral is understood in the sense of principal value. In this chapter we shall use the following notations:

$$\|\varphi\|_{p,\rho} = \left(\int_\Gamma |\varphi(t)|^p \rho(t)|dt|\right)^{1/p}$$

designates the norm in the space $L_p(\Gamma, \rho)$; $\Gamma_0 = \{z \in \mathbb{C} : |z| = 1\}$; $S_0 = S_{\Gamma_0}$; $\|\varphi\|_p = \|\varphi\|_{p,1}$; if $A \in \mathcal{L}(E)$, then we define the quotient norm of the operator A by

$$|A| = \inf_{T \in \mathcal{T}(E)} \|A + T\| \ ;$$

the indices p, ρ in the notations $\|A\|_{p,\rho}$ and $|A|_{p,\rho}$ will indicate that the underlying space in which A acts is $L_p(\Gamma, \rho)$. In 1927 M. Riesz established the boundedness of the operator S_0 in the spaces $L_p(\Gamma_0)$ with $1 < p < \infty$ [66]. This result was followed by numerous works in which it is proved that the operator S is bounded in $L_p(\Gamma, \rho)$ under various constraints on the contour Γ and the weight ρ. A survey of the literature on this problem is given, for example, in [34], [84].

It then turned out that an important role in various questions is played by the exact constants in boundedness theorems for the operator S. The first steps in this direction were made in [50(1)] and [50(4)]. Specifically, it was established that $\|S_0\| \geq \nu(p)$, where

$$\nu(p) = \begin{cases} \cot\dfrac{\pi}{2p} & \text{if} \quad 2 \leq p < \infty, \\[2ex] \tan\dfrac{\pi}{2p} & \text{if} \quad 1 < p < 2, \end{cases}$$

and that $\|S_0\|_p = \nu(p)$ for $p = 2^n$ and $p = 2^n(2^n - 1)^{-1}$. Later, the norm of the operator C:

$$(C\varphi)(t) = \frac{1}{2\pi} \int_0^{2\pi} \varphi(y) \cot\frac{y-t}{2} dt$$

acting in $L_p(0, 2\pi)$ was calculated in [63], with the result that $\|C\|_p = \nu(p)$. Subsequently, efforts were made to calculate the norms and quotient norms of singular operators in spaces $L_p(\Gamma, \rho)$. A number of results obtained in this direction are presented in this chapter.

4. SOME ESTIMATES OF NORMS AND QUOTIENT NORMS OF SINGULAR INTEGRAL OPERATORS

1. Let Γ be a simple closed Lyapunov contour, $t_1, \ldots, t_n$ distinct points on Γ, and

$$\rho(t) = \prod_{k=1}^{n} |t - t_k|^{\beta_k} (-1 < \beta_k < p - 1, \ 1 < p < \infty) . \tag{4.1}$$

In [33] it is shown that S is bounded in the space $L_p(\Gamma, \rho)$ with weight (4.1). The conditions $-1 < \beta_k < p - 1$ are also necessary for the boundedness of S in $L_p(\Gamma, \rho)$. The simplest way to prove this is by using the following simple assertion; $p^{-1} + q^{-1} = 1$.

LEMMA 4.1. *If* $S \in \mathcal{L}(L_p(\Gamma, \rho))$, *then* $\rho \in L_1(\Gamma)$ *and* $\rho^{-1} \in L_{q-1}(\Gamma)$.

PROOF. It follows from the boundedness of S in $L_p(\Gamma)$ that the operator $K = \pi i \rho^{1/p}(SR - RS)\rho^{-1/p}$, where $(R\varphi)(t) = t\varphi(t)$, is bounded in $L_p(\Gamma)$. But

$$(K\varphi)(t) = \rho^{1/p}(t) \int_\Gamma \rho^{-1/p}(\tau)\varphi(\tau)d\tau ,$$

and hence $\rho^{1/p} \in L_p$ and $\rho^{-1/p} \in L_q$, i.e., $\rho^{-1} \in L_{q-1}$. $\square$

THEOREM 4.1. *Let* $a, b \in \mathbb{C}$. *Then the following estimate holds for the operator* $A = aI + bS$ *in* $L_p(\Gamma)$:

$$|aI + bS|_p \geq \left\{ |b|^2 cot^2 \frac{\pi}{p} + \left(\frac{|a+b| - |a-b|}{2} \right)^2 \right\}^{1/2}$$

$$+ \left\{ |b|^2 cot^2 \frac{\pi}{p} + \left(\frac{|a+b| + |a-b|}{2} \right)^2 \right\}^{1/2} . \tag{4.2}$$

PROOF. Suppose first that $p = 2$ and $A = aI + bS$. Then $|A^*A| = |a + b|^2 P + |a - b|^2 Q$, $\sigma(A^*A) = \sigma(\widehat{A^*A}) = \{|a + b|^2, |a - b|^2\}$, and hence $\|A\|_2 = |A|_2 = max(|a - b|, |a + b|)$, which is in agreement with (4.2).

Now suppose $p \neq 2$. Let $A_0 = \alpha P + Q$ (with $\alpha \in \mathbb{C}$, $\alpha \neq 1$), and let

$$2\delta = \{|\alpha - 1|^2 cot^2 \frac{\pi}{p} + (|\alpha| - 1)^2\}^{1/2} + \{|\alpha - 1|^2 cot^2 \frac{\pi}{p} + (|\alpha| + 1)\}^{1/2} . \tag{4.3}$$

Consider in the plane $\mathbb{C}$ the circle γ: $|z - z_0| = R$, where $z_0 = (\delta^2 - \alpha)(\delta^2 - 1)^{-1}$ and $R = \delta |\alpha - 1| (\delta^2 - 1)^{-1}$. It is chosen so that $|z - \alpha| |z - 1|^{-1} = \delta$ for every point $z \in \gamma$. Since $\delta > |\alpha|$, it follows that $|z_0| > R$. Draw through the origin of coordinates the two tangents to the circle γ and denote by z_1 and z_2 the tangency points. The choice of the number δ guarantees that the tangents make an angle of $2\pi/\bar{p}$, where

$$\bar{p} = max(p, q) \quad (p^{-1} + q^{-1} = 1) .$$

Now consider the operator $A_1 = gP + Q$, where g is a piecewise-continuous function assuming on Γ the two values z_1 and z_2. This function is not p-nonsingular (see Sec.10); consequently, A_1 is not a Φ-operator in $L_p(\Gamma)$. We express it as

$$A_1 = \frac{\alpha - g}{\alpha - 1} \left(I + \frac{\alpha - 1}{\alpha - g} A_0 \right) .$$

Since $A_1 \notin \Phi(\mathcal{L})$,

$$\left| \frac{g - 1}{\alpha - g} A_0 \right| \geq 1 .$$

But $|\alpha - g(t)||1 - g(t)|^{-1} = \delta$, and hence $|A_0| \geq \delta$. In view of the equalities $|aI + bS|_p = |a - b||\alpha P + Q|_p$ $(\alpha = |a + b||a - b|^{-1})$ and (4.3), this yields (4.2).

$\square$

In the spaces $L_p(\Gamma, \rho)$ with the weight (4.1) one has the estimate

$$
|aI + bS|_{p,\rho} \geq \max_{k=0,1,\ldots,n} \left[\left\{ |b|^2 \cot^2 \frac{\pi(1 + \beta_k)}{p} + \left(\frac{|a + b| - |a - b|}{2} \right)^2 \right\}^{1/2} \right.
$$
$$
\left. + \left\{ |b|^2 \cot^2 \frac{\pi(1 + \beta_k)}{p} + \left(\frac{|a + b| + |a - b|}{2} \right)^2 \right\}^{1/2} \right]
\tag{4.4}
$$

(where $\beta_0 = 0$), which can be established in the same manner as in the case $\rho \equiv 1$; one has only to take care that the function g be discontinuous at the corresponding point t_k and choose in the right way the values $g(t_k \pm 0)$.

Theorem 4.1 yields the following estimates for the norms of the operator of singular integration and of the projectors $P = \frac{1}{2}(I + S)$ and $Q = \frac{1}{2}(I - S)$:

$$
|S|_p \geq \nu(p)
\tag{4.5}
$$

and

$$
|P|_p \geq \frac{1}{\sin \frac{\pi}{p}}, \quad |Q|_p \geq \frac{1}{\sin \frac{\pi}{p}}.
\tag{4.6}
$$

We shall prove below that the estimate (4.5) is exact. It is still not known whether the estimates (4.6) are exact for $p \neq 2$.

For $p = 2$ we have the following result.

THEOREM 4.2. *Let* $a, b \in \mathbb{C}$, $\Gamma = \Gamma_0$, *and* $\rho(t) = |t - 1|^\beta$ *(with* $-1 < \beta < 1$*).* *Then*

$$
|aI + bS_0|_{2,\rho} = \left\{ |b|^2 \tan^2 \frac{\pi\beta}{2} + \left(\frac{|a + b| - |a - b|}{2} \right)^2 \right\}^{1/2}
$$
$$
+ \left\{ |b|^2 \tan^2 \frac{\pi\beta}{2} + \left(\frac{|a + b| + |a - b|}{2} \right)^2 \right\}^{1/2} .
\tag{4.7}
$$

PROOF. Write the operator $A = aI + bS_0$ as $A = (a-b)(hP+Q)$, where $h = (a+b)(a-b)^{-1}$ (one can assume that $a^2 \neq b^2$). Set $A_0 = hP + Q$. The operator P is not self-adjoint for $\beta \neq 0$. It is readily seen that $P^* = \rho^{-1}P\rho$ and one has the representation $P^* = cf_+Pf_-c$, where the functions $f_+(t) = (t-1)^{-\beta}$, $f_-(t) = (1-t^{-1})^\beta$, and $c(t) = t^{\beta/2}$ are continuous at every point $t \in \Gamma_0\backslash\{1\}$. Let us show that P^* belongs to the algebra $\mathcal{A}$ generated by the singular integral operators with piecewise-continuous coefficients. Consider the operators $B = f_+Pf_-+Q$ and $R = Pt^\beta P+Q = Pf_+^{-1}f_-^{-1}P+Q$, and check directly that $B = R^{-1}$. In [50(4)] (see also Chap. V of this monograph) it is shown that the algebra $\mathcal{A}$ contains the regularizers of all its Φ-operators, and hence $B \in \mathcal{A}$. Consequently, $P^* \in \mathcal{A}$ and its symbol coincides with that of the operator $D = c(R^{-1}-Q)c$ (see Chap. V). Algebra $\mathcal{A}$ contains also the operator $M = A_0^*A_0 - \lambda I = [(\bar{h}-1)P^* + I][(h-1)P+I] - \lambda I$. A straightforward calculation shows that the value of the symbol of M at the point $z = 1$ is

$$\mathcal{M}(\mu) = \begin{bmatrix} h - \lambda + u_{11}(\mu) & u_{12}(\mu) \\ u_{21}(\mu) & 1 - \lambda + u_{22}(\mu) \end{bmatrix},$$

where

$$u_{11}(\mu) = \frac{h(\bar{h}-1)(\xi + e^{\pi i\beta}(1-\xi))^2}{\xi + e^{2\pi i\beta}(1-\xi)},$$

$$u_{22}(\mu) = \frac{(\bar{h}-1)\xi(1-\xi)(1-e^{\pi i\beta})^2}{\xi + e^{2\pi i\beta}(1-\xi)},$$

$$u_{21}(\mu)u_{12}(\mu) = u_{11}(\mu)u_{22}(\mu),$$

and

$$\xi(\mu) = \frac{sin(\pi\beta\mu)}{sin(\pi\beta)}e^{\pi i\beta(1-\mu)}.$$

The essential spectrum of the operator $A_0^*A_0$(i.e., the set of the points λ for which $A_0^*A_0 - \lambda I \notin \Phi(\mathcal{L})$) coincides with the set of those points $\lambda \in \mathbb{C}$ for which $det\,\mathcal{M}(\mu)$ vanishes for some $\mu \in [0,1]$. Consequently, $|A_0|^2$ coincides with the maximal value of the larger root of the equation $det\,\mathcal{M}(\mu) = 0$. An elementary computation shows that
$det\,\mathcal{M}(\mu) =$

$$= \lambda^2 - \lambda\left(|1-h|^2\frac{sin(\pi\beta(\mu))sin(\pi\beta(1-\mu))}{cos^2(\pi\beta/2)} + 1 + |h|^2\right) + |h|^2.$$

The largest root of this trinomial takes its maximal value (as a function of μ) for $\mu = \frac{1}{2}$ and this value is

$$\lambda_{max} = \frac{1}{4}\left\{\left(|1 - h|^2 tan^2\frac{\pi\beta}{2} + (|h| - 1)^2\right)^{1/2} + \right.$$
$$\left. + \frac{1}{4}\left(|1 - h|^2 tan^2\frac{\pi\beta}{2} + (|h| + 1)^2\right)\right\}.$$

Recalling that $A = (a - b)A_0$ and $h = (a + b)(a - b)^{-1}$, we get (4.7).

$\square$

Formula (4.7) can be extended to more general contours and weights, as we shall see in one of the subsections that follows.

2. We now prove two general results on norms of linear operators.

THEOREM 4.3. *Let $A \in \mathcal{L}(E)$ and suppose that there is a sequence $B_n \in \mathcal{L}(E)$ such that*

1) $\|B_n\varphi\| = \varphi$ *for all $\varphi \in E$,*

2) $AB_n = B_nA$ *for all $n \in I\!N$,*

and

3) $\{B_n\}$ *converges weakly to zero.*

Then $|A| = \|A\|$.

PROOF. It suffices to show that

$$\|A + T\| \geq \|A\| \tag{4.8}$$

for every compact operator T.

Pick an arbitrary $\varepsilon > 0$ and a vector $\varphi \in E$ such that $\|\varphi\| = 1$ and $\|A\varphi\| > \|A\| - \varepsilon$. Set $\varphi_n = B_n\varphi$. Since $B_n \longrightarrow 0$ weakly, it follows that $\|TB_n\varphi\| \longrightarrow 0$. Choose n so that $\|TB_n\varphi\| < \varepsilon$. Then

$$\|(A + T)\varphi_n\| \geq \|AB_n\varphi\| - \varepsilon = \|A\varphi\| - \varepsilon \geq (\|A\| - 2\varepsilon)\|\varphi_n\| ,$$

and hence $\|A + T\| \geq \|A\|$.

$\square$

We next consider two examples.

Example 4.1. Let $a, b \in \mathbb{C}$, $A = aI + bS_0$, and $(B_n\varphi)(t) = t^n\varphi(t^{2n})$. It is readily checked that the operator A and the sequence $\{B_n\}$ satisfy the conditions of Theorem 4.3 in each space $L_p(\Gamma_0)$ with $1 < p < \infty$. Consequently,

$$\|aI + bS_0\|_p = |aI + bS_0|_p \ . \tag{4.9}$$

Example 4.2. Let $\Gamma = \mathbb{R}$, $\rho(t) = t^\beta$ (with $-1 < \beta < p - 1$), $a, b \in \mathbb{C}$, $A = aI + bS_\mathbb{R}$, and $(B_n\varphi)(t) = n^{(1+\beta)/p}\varphi(nt)$. Then A and the sequence $\{B_n\}$ satisfy the conditions of Theorem 4.3, and hence

$$\|aI + bS_\mathbb{R}\|_{p,\rho} = |aI + bS_\mathbb{R}|_{p,\rho} \ . \tag{4.10}$$

THEOREM 4.4. *Let $E = L_p(M, d\mu)$ (over the real field), $A \in \mathcal{L}(E)$, and let $\tilde{A}$ denote the complexification of A:*

$$\tilde{A}f = A(Re f) + iA(Im \ f) \ .$$

Then $\tilde{A}$ is bounded in the complex space $L_p(M, d\mu)$ and $\|\tilde{A}\| = \|A\|$.

PROOF. Let $f_1 = Re \ f$, $f_2 = Im \ f$, $g_1 = Af_1$, and $g_2 = Af_2$. Integrating the inequality

$$\|g_1 cos \ \alpha + g_2 sin \ \alpha\|_p^p \leq \|A\|^p \|f_1 cos \ \alpha + f_2 sin \ \alpha\|_p^p$$

(where $0 \leq \alpha \leq 2\pi$), we get

$$\int_0^{2\pi} d\alpha \int_M |g_1(t)cos \ \alpha + g_2(t)sin \ \alpha|^p d\mu$$
$$\leq \|A\|^p \int_0^{2\pi} d\alpha \int_M |f_1(t)cos \ \alpha + f_2(t)sin \ \alpha|^p d\mu \ .$$

Permuting the order of integration we get

$$\int_M |g_1^2(t) + g_2^2(t)|^{p/2} d\mu \int_0^{2\pi} |sin(\alpha + \beta(t))|^p d\alpha$$

$$\leq \|A\|^p \int_M |f_1^2(t) + f_2^2(t)|^{p/2} d\mu \int_0^{2\pi} |sin(\alpha + \gamma(t))|^p d\alpha .$$

Since the inner integrals are independent of t, they can be cancelled. We thus get

$$\int_M |\tilde{A}f|^p d\mu \leq \|A\|^p \int_M |f|^p d\mu .$$

$\square$

3. Here we calculate the norms $\|S_0\|$ and $|S_0|$. The estimate $\|S_0\|_p \leq cot\frac{\pi}{2p}$ in the space $L_p(\Gamma_0)$ with $p = 2^n$ is readily established by induction on n with the help of the inequality

$$\|S_0\|_{2p} \leq \|S_0\|_p + \sqrt{1 + \|S_0\|_p^2} ,$$

which in turn is a consequence of the known identity

$$2S_0(\varphi S_0 \varphi) = (S_0 \varphi)^2 + \varphi^2$$

(for more details see [50(9)]).

Considerably more effort is necessary for establishing the estimate of $\|S_0\|_p$ for the intermediary values of p. We precede this by a number of auxiliary results.

LEMMA 4.2. *For* $|x| \leq \frac{\pi}{2}$ *and* $1 < p \leq 2$

$$|sin\ x|^p \leq tan^p \frac{\pi}{2p} cos^p x - \beta(p)cos\ px , \tag{4.11}$$

where

$$\beta(p) = \frac{sin^{p-1}(\pi/2p)}{cos(\pi/2p)} .$$

PROOF. Consider the function

$$F(x) = \frac{sin^p + \beta(p)cos\ px}{cos^p x}$$

on the interval $0 < x < \pi/2$. It is readily checked that

$$F'(x) = p\frac{\sin^{p-1}x}{\cos^{p+1}x}g(x) \,,$$

where

$$g(x) = 1 - \beta(p)\frac{\sin(p-1)x}{\sin^{p-1}x} \,.$$

Since $g'(x) = -\beta(p)(p-1)\sin[(2-p)x]\sin^{-p}x < 0$, it follows that g is a decreasing function. Moreover, $g(\pi/2p) = 0$, and hence $F'(x) > 0$ for $0 < x < \pi/2p$, while $F'(x) < 0$ for $\frac{\pi}{2p} < x < \frac{\pi}{2}$. Therefore, $F(x) \le F(\frac{\pi}{2p}) = \tan^p\frac{\pi}{2p}$ $(0 < x < \frac{\pi}{2})$, which yields (4.11).

$\square$

LEMMA 4.3. *The function* $g_0 : \ \ \mathbb{C} \longrightarrow \mathbb{R}$ *defined by*

$$g_0(z) = \begin{cases} 0 \,, & z = 0 \\ |z|^p\cos(p\alpha(z)) \,, & z \ne 0 \,, \end{cases}$$

where $\alpha(x + iy) = \arctan(y/|x|)$ *and* $1 < p \le 2$, *is subharmonic in* $\mathbb{C}$.

PROOF. We have to show that the inequality

$$g_0(z_0) \le \frac{1}{2\pi}\int_0^{2\pi} g_0(z_0 + re^{i\theta})d\theta \tag{4.12}$$

holds for every $z_0 \in \mathbb{C}$ and every $r > 0$. It is known that in order that a continuous real-valued function be subharmonic it is necessary and sufficient that the inequality (4.12) be fulfilled for every z_0 and small enough values of r.

In the right (left) half plane $Re\ z > 0$ (respectively, $Re\ z < 0$), g_0 coincides with the harmonic function $Re\ z^p$, with $|arg\ z| < \frac{\pi}{2}$ (respectively, $Re(-z)^p$, with $|arg(-z)| < \frac{\pi}{2}$). It follows that for each z with $Re\ z \ne 0$, inequality (4.12) holds for sufficiently small values of r.

Now let $z = 0$. Then

$$\frac{1}{2\pi}\int_{-\pi}^{\pi} g_0(re^{i\theta})d\theta = \frac{1}{\pi}\int_{-\pi/2}^{\pi/2} r^p\cos(p\theta)d\theta$$

$$= \frac{2r^p}{\pi p}\sin\frac{p\pi}{2} \ge 0 = g_0(0)$$

for all $r > 0$, i.e., (4.12) holds for $z = 0$, too.

Finally, let $z = iy$, with $y \in \mathbb{R} \setminus \{0\}$. Let $h_0(z) = Re \; z^p$, where $z \neq 0$ and $|arg| < \pi$. The functions h_0 and g_0 agree in the right half-plane $Re \; z \geq 0$. We show that in the left half plane $Re \; z < 0$ the difference $g_0(z) - h_0(z)$ is nonnegative at every point z with $Im \; z \neq 0$. If $z = ae^{ix}$ (with $a > 0$ and $\frac{\pi}{2} < x < \pi$), then

$$g_0(z) - h_0(z) = a^p[cos \; p(x - \pi) - cos \; px]$$
$$= 2a^p sin(p(x - \frac{\pi}{2}))sin\frac{\pi p}{2} \geq 0 \; ,$$

and if $z = ae^{ix}$ (with $a < 0$ and $-\pi < x < -\frac{\pi}{2}$), then

$$g_0(z) - h_0(z) = a^p(cos \; p(x + \pi) - cos \; px)$$
$$= 2a^p sin(p(x + \frac{\pi}{2}))sin\frac{\pi p}{2} \geq 0 \; .$$

But the function h_0 is harmonic in $\mathbb{C} \setminus \{z : \; Re \; z < 0, \; Im \; z = 0\}$, so that

$$g_0(iy) = h_0(iy) = \frac{1}{2\pi} \int_{-\pi}^{\pi} h_0(iy + re^{i\theta})d\theta$$
$$\leq \frac{1}{2\pi} \int_{-\pi}^{\pi} g_0(iy + re^{i\theta})d\theta$$

for every r, $0 < r < |y|$.

$\square$

LEMMA 4.4. *Let*

$$(K_0 f)(t) = \frac{1}{2\pi} \int_{\Gamma_0} f(t)|dt|$$

and $1 < p \leq 2$. *Then*

$$\|S_0 - K_0\|_p \leq tan\frac{\pi}{2p} \; . \tag{4.13}$$

PROOF. The operator $i(S_0 - K_0)$ takes real-valued functions again into such functions (this is readily verified on trigonometric polynomials). By Theorem 4.4, it suffices to prove (4.13) in the L_p-space of real-valued functions. Thus, it suffices to show that for $1 < p \leq 2$

$$\|g\|_{L_p(\Gamma_0)} \leq tan\frac{\pi}{2p}\|f\|_{L_p(\Gamma_0)} , \tag{4.14}$$

where

$$f(t) = \sum_{k=-n}^{n} f_k t^k , \quad g(t) = i\left(\sum_{k=-n}^{-1} - \sum_{k=1}^{n}\right) f_k t^k ,$$

with $|t| = 1$, $f_k \in \mathbb{C}$, and $f_{-k} = \bar{f}_k$.

Notice that $f + ig = h$, where

$$h(t) = 2\sum_{k=1}^{n} f_k t^k + f_0 .$$

We set $w(z) = g_0(h(z))$, where g_0 is the subharmonic function considered in Lemma 4.3. Since h is holomorphic in $\mathbb{C}$, w is also subharmonic in $\mathbb{C}$. Let $\alpha(x + iy) = arctan(y/|x|)$ and $\psi(z) = \alpha(h(z))$. The function ψ is defined at all points z for which $h(z) \neq 0$. We extend it arbitrarily at the roots of the polynomial h, which allows us to express the function w as $w = |h|^p cos(p\psi)$. It is easily verified that for $|z| = 1$,

$$|f(z)| = |h(z)|cos\ \psi(z) \ \text{and}\ g(z) = |h(z)|sin\ \psi(z) . \tag{4.15}$$

Next, it follows from Lemma 4.2 that

$$\begin{aligned}\int_{\Gamma_0} |h(z)|^p |sin\ \psi(z)|^p |dz| \leq & tan^p\frac{\pi}{2p}\int_{\Gamma_0} |h(z)|^p |cos^p\psi(z)||dz| \\ & - \beta(p)\int_{\Gamma_0} |h(z)|^p cos(p\psi(z))|dz| .\end{aligned} \tag{4.16}$$

Let us show that the last integral in the right-hand side of this inequality is nonnegative.

Since f is a real-valued function, $Im\ h(0) = 0$. Consequently, if $h(0) \neq 0$, then $\psi(0) = 0$, and hence $w(0) = |h(0)|^p$. Since w is subharmonic

$$\begin{aligned}0 \leq w(0) \leq & \frac{1}{2\pi}\int_0^{2\pi} w(e^{i\theta})d\theta \\ = & \frac{1}{2\pi}\int_{\Gamma_0} |h(z)|^p cos(p\psi(z))|dz| .\end{aligned} \tag{4.17}$$

Relations (4.15)-(4.17) yield (4.14).

$\qquad\qquad\qquad\qquad\qquad\qquad\qquad\qquad\qquad\qquad\qquad\qquad\qquad$ □

THEOREM 4.5. *For every* p, $1 < p < \infty$,

$$\|S_0\|_p = |S_0|_p = \nu(p) = \begin{cases} \cot\frac{\pi}{2p} & \text{if} \quad 2 \le p < \infty \\ \tan\frac{\pi}{2p} & \text{if} \quad 1 < p \le 2 . \end{cases} \qquad (4.18)$$

PROOF. For $1 < p \le 2$ the equality $|S_0|_p = \nu(p)$ follows from (4.5) and (4.13). For $p > 2$ it is established by passing to the conjugate operator. The equality $\|S_0\|_p = |S_0|_p$ has been proven above (see (4.9)).

$\qquad\qquad\qquad\qquad\qquad\qquad\qquad\qquad\qquad\qquad\qquad\qquad\qquad$ □

Remark 4.1. It follows from the relations (4.5) and (4.13) that $\|S_0 - K_0\|_p = \nu(p)$. Let $\xi : L_p(\Gamma_0) \longrightarrow L_p(0, 2\pi)$ denote the isometric mapping defined by $(\xi\varphi)(x) = \varphi(e^{ix})$. A straightforward calculation shows that $i(K_0 - S_0) = \xi C \xi^{-1}$, where

$$(C\varphi)(x) = \frac{1}{2\pi} \int_0^{2\pi} \cot\frac{t-x}{2}\varphi(t)dt . \qquad (4.19)$$

Therefore,

$$|C|_p = \|C\|_p = \nu(p) . \qquad (4.20)$$

Remark 4.2. In Zygmund's monograph [81] it is proved that $\|S_{\mathbb{R}}\|_p \le \|C\|_p$, where C is the operator defined by formula (4.19). It follows from (4.20) that $\|S_{\mathbb{R}}\| \le \nu(p)$. Relation (4.5) is also valid for $\Gamma = \mathbb{R}$ (see the proof of (4.5)). Thus,

$$|S_{\mathbb{R}}|_p = \|S_{\mathbb{R}}\|_p = \nu(p) . \qquad (4.21)$$

4. We need a number of results on the norms of singular integral operators.

THEOREM 4.6. *Let* $t_1, \ldots, t_n$ *be distinct points of* $\mathbb{R}$,

$$\tau_k = (t_k + i)(t_k - i)^{-1} , \quad k = 1, \ldots, n ,$$

$$\tau_0 = 1 ,$$

and let $p, \alpha, \alpha_1, \ldots, \alpha_n$ be real numbers which satisfy the conditions

$$\left. \begin{array}{c} 1 < p < \infty, \quad -1 < \alpha_k < p-1, \quad k = 1, \ldots, n \\ -1 < \alpha + \sum_{k=1}^{n} \alpha_k < p-1 \end{array} \right\} . \tag{4.22}$$

Set

$$\rho_1(t) = (t^2 + 1)^{\alpha/2} \prod_{k=1}^{n} |t - t_k|^{\alpha_k}$$

and

$$\rho_2(t) = \prod_{k=0}^{n} |\tau - \tau_k|^{\alpha_k} ,$$

where

$$\alpha_0 = p - 2 - \alpha - \sum_{k=1}^{n} \alpha_k .$$

Then

$$\|S_{\mathbb{R}}\|_{p,\rho_1} = \|S_0\|_{p,\rho_2} , \quad |S_{\mathbb{R}}|_{p,\rho_1} = |S_0|_{p,\rho_2} . \tag{4.23}$$

PROOF. Set

$$(B\varphi)(t) = \frac{i}{t-1} \varphi(i\frac{t+1}{t-1}) .$$

It is readily checked that

$$S_{\mathbb{R}} = -B^{-1} S_0 B$$

and

$$\| B\varphi \|_{L_p(\Gamma_0, \rho_2)} = \delta \|\varphi\|_{L_p(\mathbb{R}, \rho_1)} ,$$

where the constant δ depends on the L_p-spaces, but not on φ. This yields (4.23).

$\square$

COROLLARY 4.1. Let $p \geq 2$ and $\rho(t) = |t + i|^{p-2}$. Then

$$\|S_{\mathbb{R}}\|_{p,\rho} = |S_{\mathbb{R}}|_{p,\rho} = \nu(p) . \tag{4.24}$$

PROOF. It suffices to put $\alpha = p - 2$ and $\alpha_k = 0$, $k = 0, 1, \ldots, n$, in (4.22), and use equalities (4.19) and (4.23).

$\square$

COROLLARY 4.2. Let $p \geq 2$ and $\rho(t) = |t - 1|^{p-2}$. Then

$$\|S_0\|_{p,\rho} = |S_0|_{p,\rho} = \nu(p) \ . \tag{4.25}$$

PROOF. Put $\alpha_0 = p - 2$ and $\alpha = \alpha_1 = \ldots = \alpha_n = 0$ in (4.22) and use equalities (4.21) and (4.23).

$\square$

THEOREM 4.7. Let Γ be a simple Lyapunov contour. Define the norm in the space $L_p(\Gamma, \rho)$ by

$$\|\varphi\|_{p,\rho} = \left\{ \int_\Gamma |\varphi(t)|^p \rho(t) |dt| \right\}^{1/p} \ , \quad 1 < p < \infty \ .$$

Then

$$\|S_\Gamma\|_{p,\rho} = \|S_\Gamma\|_{q,\rho^{1-q}} \ , \qquad |S_\Gamma|_{p,\rho} = |S_\Gamma|_{q,\rho^{1-q}} \ , \tag{4.26}$$

where $p^{-1} + q^{-1} = 1$.

PROOF. The spaces $L_p(\Gamma, \rho)$ and $L_q(\Gamma, \rho^{1-q})$ are mutually conjugate. Let F_Γ denote the operator acting in $L_p(\Gamma, \rho)$ according to the rule

$$(F_\Gamma \varphi)(t) = \overline{f_\Gamma(t)\varphi(t)} \ ,$$

where $f_\Gamma(t) = exp(i\theta_\Gamma)$ and $\theta_\Gamma(t)$ denotes the angle that the tangent to Γ at the point t makes with the positively oriented x-axis. It is readily checked that $S_\Gamma^* = -F_\Gamma S_\Gamma F_\Gamma$, which in turn yields (4.26).

$\square$

COROLLARY 4.3. Let $1 < p \leq 2$ and $\rho(t) = |t + i|^{2-p}$. Then

$$\|S_{\mathrm{IR}}\|_{p,\rho} = |S_{\mathrm{IR}}|_{p,\rho} = \nu(p) \ . \tag{4.27}$$

PROOF. Use Corollary 4.1 and Theorem 4.7.

In a similar manner one proves

COROLLARY 4.4. *Let* $1 < p \leq 2$ *and* $\rho(t) = |t-1|^{2-p}$. *Then*

$$\|S_0\|_{p,\rho} = |S_0|_{p,\rho} = \nu(p) .$$ (4.28)

□

Remark 4.3. We established the equalities of the type $\|S\| = |S|$ for the circle and real line on purpose. As it turns out, if Γ is a simple closed Lyapunov contour and $\|S_\Gamma\| = |S_\Gamma|$, then Γ is necessarily a circle. In fact, suppose that Γ is not a circle; then $S_\Gamma^* \neq S_\Gamma$ (see [50(10)] and [50(9)]). Since $S_\Gamma^2 \neq I$, we have $S_\Gamma S_\Gamma^* \neq I$, and hence the spectrum of the operator $S_\Gamma S_\Gamma^*$ contains a point $\lambda \neq 1$. It follows from the equality $S_\Gamma^* S_\Gamma - \lambda I = S_\Gamma^*(I - \lambda S_\Gamma^* S_\Gamma)S_\Gamma$ that the spectrum of $S_\Gamma^* S_\Gamma$ contains a point $\lambda_0 > 1$. This implies that $\|S_\Gamma\| > 1$. In Sec. 6 we shall prove that $|S_\Gamma|_2 = |S_0|_2 = 1$.

We shall also need

THEOREM 4.8. *Let* $\min(0, p-2) \leq \beta \leq \max(0, p-2)$. *Then*

$$\|S_0\|_{p,\rho} = \nu(p) \quad (\rho(t) = |1-t|^\beta) .$$ (4.29)

□

PROOF. It suffices to apply the well-known Riesz-Stein [74] interpolation theorem to the pair of spaces $L_p(\Gamma_0)$, $L_p(\Gamma_0, |\theta-1|^{p-2})$ and use equalities (4.19), (4.25), (4.28) and the estimate (4.4).

□

Since in this chapter we shall resort several times to the Riesz-Stein theorem, we state it here in a form convenient for our purposes.

Let X be a space on which there is given a nonnegative, completely additive measure and let $\rho_i : X \to \mathbb{R}^+$, $i = 0, 1$, be a.e. finite functions. Set $E_i = L_{p_i}(X, \rho_i)$

$(1 \leq p_i \leq \infty)$, $i = 0, 1$, and $E_\theta = L_{p\theta}(X, \rho_\theta)$, $0 \leq \theta \leq 1$, where p_θ and ρ_θ are defined by the formulas

$$p_\theta^{-1} = (1 - \theta)p_0^{-1} + \theta p_1^{-1}$$

and

$$\rho_\theta = \rho_0^{(1-\theta)/p_0} \cdot \rho_1^{\theta/p_1} .$$

THEOREM 4.4. *If $A \in \mathcal{L}(E_i)$, $i = 0, 1$, then $A \in \mathcal{L}(E_\theta)$ for all $\theta \in [0, 1]$ and*

$$\|A\|_{E_\theta} \leq \|A\|_{E_0}^{1-\theta} \|A\|_{E_1}^{\theta} . \tag{4.30}$$

$\square$

5. EXACT CONSTANTS IN THE THEOREMS OF HARDY-LITTLEWOOD, BABENKO, AND KHVEDELIDZE ON BOUNDEDNESS OF SINGULAR INTEGRAL OPERATORS

1. We let $\nu(p, \alpha)$ denote the numbers defined for $2 \leq p < \infty$ by the formulas

$$\nu(p, \alpha) = \begin{cases} \cot\dfrac{\pi(p - 1 - \alpha)}{2p} & \text{if} \quad p - 2 < \alpha < p - 1, \\[2ex] \cot\dfrac{\pi}{2p} & \text{if} \quad 0 \leq \alpha \leq p - 2, \\[2ex] \cot\dfrac{\pi(1 + \alpha)}{2p} & \text{if} \quad -1 < \alpha < 0 , \end{cases} \tag{5.1}$$

and

$$\nu(p, \alpha) = \nu(q, \alpha(1 - q)) \quad \text{if} \quad 1 < p < 2 ,$$

where $p^{-1} + q^{-1} = 1$.

THEOREM 5.1. *Let $t_0 \in p$. Then*

$$\|S_0\|_{p, |t-t_0|^\alpha} = \nu(p, \alpha) , \quad -1 < \alpha < p - 1 . \tag{5.2}$$

PROOF. Let $2 \leq p < \infty$ and $p - 2 < \alpha < p - 1$ and set

$$r = p(p - 1 - \alpha)^{-1} \quad \text{and} \quad r' = r(r - 1)^{-1} .$$

Then $r' < p < r$, and hence we can write $1/p$ in the form $1/p = (1 - \theta)/r + \theta/r'$. Notice that $(1 - \theta)(r - 2) = \alpha(p - 1 - \alpha)^{-1} = \alpha r p^{-1}$, and hence $|t - t_0|^{\alpha/p} = |t - t_0|^{(1-\theta)(r-2)r^{-1}}$. By Theorem 4.9,

$$\|S_0\|_{p,|t-t_0|^\alpha} \leq \|S_0\|_{r,|t-t_0|^{r-2}}^{1-\theta} \|S_0\|_{r'}^\theta . \tag{5.3}$$

Since $r > 2$, it follows from formula (4.28) that

$$\|S_0\|_{r,|t-t_0|^{r-2}} = \|S_0\|_{r'} = \cot\frac{\pi}{2r} .$$

In view of (5.3), this yields the estimate

$$\|S_0\|_{p,|t-t_0|^\alpha} \leq \cot\frac{\pi}{2r} = \nu(p, \alpha) \tag{5.4}$$

for $p \geq 2$ and $p - 2 < \alpha < p - 1$. In particular, if $0 < \alpha < 1$, then

$$\|S_0\|_{2,|t-t_0|^\alpha} \leq \nu(2, \alpha) .$$

By Theorem 4.7,

$$\|S_0\|_{2,|t-t_0|^\alpha} = \|S_0\|_{2,|t-t_0|^{-\alpha}} .$$

Since $\nu(2, \alpha) = \nu(2, -\alpha)$, the estimate

$$\|S_0\|_{p,|t-t_0|^\alpha} \leq \nu(p, \alpha)$$

holds also for $p = 2$ and $-1 < \alpha < 0$.

Now let $2 < p < \infty$ and $-1 < \alpha < 0$ and set

$$r = p(1 + \alpha)^{-1} \quad \text{and} \quad \beta = (2 + 2\alpha - p)p^{-1} .$$

It is readily checked that $2 < p < r$ and $-1 < \beta < 0$. We write the number $1/p$ in the form $1/p = (1 - \theta)/2 + \theta/r$, where $\theta = (p - 2)(p - 2 - 2\alpha)^{-1} \in (0,1)$. Since $\alpha/p = \beta(1 - \theta)/2$, we can apply Theorem 4.9 once more to obtain

$$\|S_0\|_{p,|t-t_0|^\alpha} \leq \|S_0\|_{2,|t-t_0|^\beta}^{1-\theta} \|S_0\|_r^\theta . \tag{5.5}$$

As was shown above,

$$\|S_0\|_{2,|t-t_0|^\beta} \leq \nu(2,\beta) = \cot\frac{\pi(1+\alpha)}{2p} .$$

Since, in addition,

$$\|S_0\|_r = \cot\frac{\pi}{2r} = \cot\frac{\pi(1+\alpha)}{2p} ,$$

inequality (5.5) yields the estimate

$$\|S_0\|_{p,|t-t_0|^\alpha} \leq \nu(p,\alpha) \quad \text{for} \quad p > 2 \quad \text{and} \quad -1 < \alpha < 0 .$$

For values $0 \leq \alpha \leq p - 2$ an analogous estimate was obtained in Theorem 4.8. We have thus established the needed estimate for all $p \geq 2$ and $-1 < \alpha < p - 1$. By Theorem 4.7, the estimate

$$\|S_0\|_{p,|t-t_0|^\alpha} \leq \nu(p,\alpha) \tag{5.6}$$

is valid also for $1 < p < 2$. To complete the proof of the theorem it remains to use inequality (4.4).

$\square$

Proceeding in much the same way one can prove the following result:

THEOREM 5.2. *Let* $1 < p < \infty$, $-1 < \alpha < p - 1$, *and* $t_0 \in \mathbb{R}$. *Then*

$$\|S_\mathbb{R}\|_{p,|t-t_0|^\alpha} = \nu(p,\alpha) . \tag{5.7}$$

$\square$

2.   It follows from inequality(5.2) that the norm of the operator S_0 does not depend on the point $t_0 \in \Gamma$ which defines the weight $|t - t_0|^\alpha$. What can be said in the case of the weights $\Pi_{k=1}^n |t - t_k|^{a_k}$? It turns out that for $n \geq 2$ the norm depends on the relative positioning of the points $t_1, \ldots, t_n$.

Suppose that the numbers $\alpha_1, \ldots, \alpha_n$ satisfy the inequalities

$$-1 < \alpha_j < 0 \quad \text{for} \quad j = 1, \ldots, m$$

and

$$0 \leq \alpha_j < p - 1 \quad \text{for} \quad j = m+1, \ldots, n \ ,$$

where $m \leq n$. We put

$$\beta_1 = \sum_{j=1}^m \alpha_j \ , \quad \beta_2 = \sum_{j=m+1}^n \alpha_j \ ,$$

and

$$\bar{\alpha} = \begin{cases} \beta_1 & \text{if} \quad \beta_1 + \beta_2 < p - 2 \ , \\ \beta_2 & \text{if} \quad \beta_1 + \beta_2 \geq p - 2 \ . \end{cases}$$

THEOREM 5.3.   *Let*

$$\rho(t) = \prod_{k=1}^n |t - t_k|^{\alpha_k}$$

where $t_k \in \Gamma_0$ and $t_k \neq t_j$ for $k \neq j$. If $\bar{\alpha} \in (-1, p-1)$, then

$$\sup_{t_1, \ldots, t_n \in \Gamma_0} \|S_0\|_{p,\rho} = \nu(p, \bar{\alpha}) \ , \tag{5.8}$$

whereas if $\bar{\alpha} \notin (-1, p-1)$,

$$\sup_{t_1, \ldots, t_n \in \Gamma_0} \|S_0\|_{p,\rho} = \infty \ . \tag{5.9}$$

PROOF. By Theorem 4.7, it suffices to examine the case where $\beta_1 + \beta_2 \geq p - 2$. In fact, if $\beta_1 + \beta_2 < p - 2$, then $(\beta_1 + \beta_2)(1 - q) > (p - 2)(1 - q) = q - 2$. Then $\bar{\alpha} = \beta_2$.

First let us establish equality (5.9). Suppose that it is not valid, i.e., that $\bar{\alpha} \notin (-1, p-1)$, but $\sup \|S_0\|_{p,\rho} = c_0 < \infty$. Then

$$\|S_0 f\|_{p,\rho} \leq c_0 \|f\|_{p,\rho} \tag{5.10}$$

for every trigonometric polynomial $f = \Sigma f_k t^k$, where the constant c_0 does not depend upon the choice of the points $t_1, \ldots, t_n \in \Gamma_0$. Now letting $t_{m+1} \to t_n, \ldots, t_{n-1} \to t_n$ in (5.10), we get

$$\|S_0\|_{p,\rho_1} \leq c_0 \|f\|_{p,\rho_1}, \tag{5.11}$$

where

$$\rho_1(t) = |t - t_n|^{\beta_2} \prod_{k=1}^{m} |t - t_k|^{\alpha_k}.$$

Since the set of trigonometric polynomials is dense in $L_p(\Gamma_0, \rho_1)$ (see, e.g., [50(9)]), it follows from (5.11) that the operator S_0 is bounded in $L_p(\Gamma_0, \rho_1)$, which contradicts Lemma 4.1. Equality (5.9) is thus proved.

Now let $\bar{\alpha} \in (-1, p-1)$. Since $\beta_2 = \bar{\alpha}$, inequality (4.4) implies that $\|S_0\|_{p,\rho_1} \geq \nu(p, \bar{\alpha})$. Hence, for every $\varepsilon > 0$ there is a trigonometric polynomial f such that $\|S_0 f\|_{p,\rho_1} > (\nu(p, \bar{\alpha}) - \varepsilon)\|f\|_{p,\rho_1}$. Upon choosing the points $t_{m+1}, \ldots, t_{n-1}$ sufficiently close to t_n, we get

$$\|S_0 f\|_{p,\rho_1} > (\nu(p, \bar{\alpha}) - \varepsilon)\|f\|_{p,\rho_1},$$

whence

$$\sup_{t_1,\ldots,t_n \in \Gamma_0} \|S_0\|_{p,\rho} \geq \nu(p, \bar{\alpha}).$$

Now let us show that $\|S_0\|_{p,\rho} \leq \nu(p, \bar{\alpha})$ for every choice of the points $t_1, \ldots, t_n$. Suppose first that $n = 2$, $\beta_1 \leq 0$, $\beta_2 \geq 0$, and $\beta_1 + \beta_2 \geq p - 2$. It follows from Theorems 5.2 and 4.6 that

$$\|S_0\|_{p,\rho_2} = \|S_{\mathbb{R}}\|_{\rho,|t|^{\beta_2}} = \nu(p, \beta_2),$$

where

$$\rho_2(t) = |t - t_1|^{-\beta_2 - 2 + p} |t - t_2|^{\beta_2}.$$

Notice that

$$w(t) = |t - t_1|^{\beta_1}|t - t_2|^{\beta_2} = (|t - t_1|^{p-2-\beta_2}|t - t_2|^{\beta_2})^{1-\theta}|t - t_2|^{\beta_2\theta}$$

with $\theta \in [0, 1]$.

It follows from Theorems 4.9 and 5.1 that

$$\|S_0\|_{p,w} \leq \|S_0\|_{p,\rho_2}^{1-\theta}\|S_0\|_{p,|t-t_2|^{\beta_2}}^{\theta} \leq \nu(p, \beta_2) = \nu(p, \bar{\alpha}) \ .$$

To derive the estimate $\|S_0\|_{p,\rho} \leq \nu(p, \bar{\alpha})$ in the general case we need the following statement which is obtained by successive application of Theorem 4.9.

LEMMA 5.1. *Let* $\lambda_1, \ldots, \lambda_n \geq 0$, $\lambda_1 + \ldots + \lambda_n \leq 1$, *and suppose that the weight* $\rho:\Gamma_0 \to \mathbb{R}^+$ *admits the representation* $\rho = \Pi_{k=1}^n \rho_k^{\lambda_k}$. *If* $\|A\|_{p,\rho_k} \leq M$, $k = 1, \ldots, n$, *then* $\|A\|_{p,\rho} \leq M$.

$\square$

We turn now to the proof of the theorem. Let $\beta_1 \neq 0$ and $\beta_2 \neq 0$. We set

$$\rho_{ij}(t) = |t - t_i|^{\beta_1}|t - t_j|^{\beta_2} \quad (i = 1, \ldots, m; \ j = m+1, \ldots, n)$$

and

$$\lambda_{ij} = \frac{\alpha_i\alpha_j}{\beta_1\beta_2} \quad (\beta_1 = \alpha_1 + \ldots + \alpha_m; \ \beta_2 = \alpha_{m+1} + \ldots + \alpha_n) \ .$$

Then $\lambda_{ij} \geq 0$, $\Sigma\lambda_{ij} = 1$, and $\Pi\rho_{ij} = \rho$. Hence, by Lemma 5.1,

$$\|S_0\|_{p,\rho} \leq \max_{i,j}\|S_0\|_{p,|t-t_i|^{\beta_1}|t-t_j|^{\beta_2}} \leq \nu(p, \beta_2) = \nu(p, \bar{\alpha}) \ .$$

The case in which one of the numbers β_1 or β_2 is equal to zero is treated analogously: one only has to take $\lambda_{ij} = \alpha_j/\beta_2$ if $\beta_1 = 0$ or $\lambda_{ij} = \alpha_j/\beta_1$ if $\beta_2 = 0$.

$\square$

From (5.8) and the estimate (4.4) we obtain the following result:

COROLLARY 5.1.   *Let $2 \leq p < \infty$ and $\rho(t) = \Pi_{k=1}^{n}|t - t_k|^{\alpha_k}$, with $t_k \neq t_j$ for $k \neq j$. Suppose one of the following conditions is satisfied:*

1) $\alpha_1 + \ldots + \alpha_n \leq p - 2$, $\alpha_k \geq 0$ $(k = 1, \ldots, n)$;

2) $\alpha_1 + \ldots + \alpha_n \leq p - 2$, $\alpha_1 \leq 0$, $\alpha_k \geq 0$ $(k = 2, \ldots, n)$;

3) $\alpha_1 + \ldots + \alpha_n \geq p - 2$, $\alpha_1 \geq p - 2$, $\alpha_k \leq 0$ $(k = 2, \ldots, n)$.

Then

$$\|S_0\|_{p,\rho} = \max_{1 \leq k \leq n} \nu(p, \alpha_k) \ . \tag{5.12}$$

The case $1 < p < 2$ reduces to the case $p \geq 2$ upon replacing p by q (where $p^{-1} + q^{-1} = 1$) and α_k by $\alpha_k(1 - q)$.

□

Theorem 5.3 and Corollary 5.1 can be extended, with the help of Theorem 4.6, to the spaces $L_p(\mathbb{R}, (t^2 + 1)^{\alpha/2}\Pi|t - t_k|^{\alpha_k})$.

We have the following supplement to Theorem 5.2.

THEOREM 5.4.   *Let $1 < p < \infty$, $-1 < \alpha < p - 1$, $-1 < \beta < p - 1$, $-1 < \alpha + \beta < p - 1$, and let $t_1, t_2 \in \mathbb{R}$, $t_1 \neq t_2$. Then*

$$\|S_{\mathbb{R}}\|_{p,|t-t_1|^{\alpha}|t-t_2|^{\beta}} = \max(\nu(p, \alpha), \nu(p, \beta), \nu(p, \alpha + \beta)) \ . \tag{5.13}$$

PROOF. By Theorem 4.6,

$$\|S_{\mathbb{R}}\|_{p,|t-t_1|^{\alpha}|t-t_2|^{\beta}} = \|S_0\|_{p,|\tau-\tau_1|^{\alpha}|\tau-\tau_2|^{\beta}|\tau-1|^{p-2-\alpha-\beta}} \ ,$$

where $\tau_1, \tau_2 \in \Gamma_0$, $\tau_1 \neq \tau_2$, $\tau_i \neq 1$ $(i = 1, 2)$. The numbers α, β and $p - 2 - \alpha - \beta$ satisfy one of the conditions 1) - 3) of Corollary 5.1. In fact, if $\alpha \leq 0$ and $\beta \leq 0$, then $p - 2 - \alpha - \beta \geq p - 2$, and hence condition 3) is satisfied. If $\alpha \geq 0$ and $\beta \geq 0$, then for

$\alpha + \beta \leq p - 2$ (respectively, $\alpha + \beta > p - 2$) condition 2) (respectively, 3)) is satisfied. Therefore, Corollary 5.1 yields

$$\|S_0\|_{p,|\tau-\tau_1|^\alpha|\tau-\tau_2|^\beta|\tau-1|^{p-2-\alpha-\beta}} = \max(\nu(p,\alpha), \nu(p,\beta), \nu(p,p-2-\alpha-\beta)) \ .$$

It remains to observe that $\nu(p, p-2-\alpha-\beta) = \nu(p, \alpha+\beta)$.

$\square$

<u>Remark 5.1.</u> In contrast to the norm of the operator S_0 in $L_p(\Gamma_0, |\tau - \tau_1|^\alpha|\tau - \tau_2|^\beta)$, the norm of the operator $S_{\mathbb{R}}$ in $L_p(\mathbb{R}, |t - t_1|^\alpha|t - t_2|^\beta)$ does not depend on the positioning of the points $t_1, t_2 \in \mathbb{R}$ provided $t_1 \neq t_2$.

THEOREM 5.5. *Let* $-1 < \alpha < 1$, $-1 < \beta < 1$, *and* $\rho(t) = |t - 1|^\alpha|t + 1|^\beta$. *Then*

$$\|S_0\|_{2,\rho} = \max(\nu(2,\alpha), \nu(2,\beta)) \ .$$

PROOF. If $\varphi(t) = \Sigma\varphi_k t^k$ is a trigonometric polynomial, then in the space $L_2(\Gamma_0, \rho_1)$ with $\rho_1(t) = |t - 1|^\alpha|t + 1|^\alpha$

$$\|\varphi\|^2 = \int_{\Gamma_0} |t^2 - 1|^\alpha |\Sigma\varphi_k t^k|^2 |dt|$$

$$= \int_{\Gamma_0} |(t^2 - 1)^{\alpha/2}\Sigma\varphi_{2k}t^{2k} + t(t^2 - 1)^{\alpha/2}\Sigma\varphi_{2k+1}t^{2k}|^2 |dt| \ .$$

The functions $f_1(t) = (t^2-1)^{\alpha/2}\Sigma\varphi_{2k}t^{2k}$ and $f_2(t) = t(t^2-1)^{\alpha/2}\Sigma\varphi_{2k+1}t^{2k}$ are orthogonal in $L_2(\Gamma_0)$, and hence

$$\|\varphi\|^2_{L_2(\Gamma_0,\rho)} = \int_{\Gamma_0} |f_1(t)|^2 |dt| + \int_{\Gamma_0} |f_2(t)|^2 |dt|$$

$$= \int_{\Gamma_0} |t - 1|^\alpha |g_1(t)|^2 |dt| + \int_{\Gamma_0} |t - 1|^\alpha |g_2(t)|^2 |dt| \ ,$$

where $g_1(t) = \Sigma\varphi_{2k}t^k$ and $g_2(t) = \Sigma\varphi_{2k+1}t^k$. The equality

$$\|S_0\varphi\|^2_{2,\rho} = \|S_0 g_1\|^2_{2,|t-1|^\alpha} + \|S_0 g_2\|^2_{2,|t-1|^\alpha}$$

is proved in a similar manner. This yields

$$\|S_0\|_{2,|t-1|^\alpha|t+1|^\alpha} = \|S_0\|_{2,|t-1|^\alpha} = \nu(2,\alpha) \ .$$

Moreover, it follows from Corollary 5.1 that

$$\|S_0\|_{2,|t-1|^\alpha|t+1|^{-\alpha}} \leq \nu(2,\alpha) \ .$$

Suppose, for the sake of definiteness, that $|\beta| \leq |\alpha|$, and represent the weight ρ in the form $|t-1|^\alpha|t+1|^\beta = (|t-1|^\alpha|t+1|^\alpha)^{1-\theta}(|t-1|^\alpha|t+1|^{-\alpha})^\theta$, where $\theta \in [0,1]$. Applying Theorem 4.9, we get

$$\|S_0\|_{2,\rho} \leq \nu(2,\alpha) = \max(\nu(2,\alpha),\nu(2,\beta)) \ .$$

The opposite inequality follows from estimate (4.4).

$$\square$$

6. THE QUOTIENT NORM OF OPERATORS OF LOCAL TYPE

We have seen in the preceding sections that the norm of the operator S_Γ depends on the shape of the contour Γ (Remark 4.3) as well as on the positioning of the points $t_1,\ldots,t_n$ that define the weight ρ (Theorem 5.3).

It turns out that the quotient norm $|S_\Gamma|$ depends neither on the form of the (Lyapunov!) contour Γ, nor on the positioning of the points $t_1,\ldots,t_n$ (provided that they are all distinct). This assertion is valid not only for S_Γ , but rather for every operator of local type, as we shall now prove.

Let X be a compact Hausdorff topological space, and let μ be a nonnegative (possibly infinite) measure defined on a σ-algebra which contains all the Borel subsets of

X. Let E be a Banach space of functions $f : X \longrightarrow \mathbb{C}$. For each measurable subset $M \subset X$ we let P_M denote the operator of multiplication by the indicator function ξ_M of M. In this section we deal only with spaces in which the operators P_M are bounded.

We remind the reader that the operator $A \in \mathcal{L}(E)$ is said to be of *local type* if the operator $P_{F_1} A P_{F_2}$ is compact for every pair of disjoint closed sets $F_1, F_2 \subset X$ [71].

Following [71] we set

$$q(x, A) = \inf_U |P_U A| \tag{6.1}$$

where the infimum is taken over all the neighborhoods U of the point x.

We say that E belongs to the class k if E is an ideal Banach space (see [86]) and

$$|P_{M_1} A P_{M_1} + P_{M_2} B P_{M_2}| \le max(|A\|, |B|) \tag{6.2}$$

for any pair of disjoint Borel sets $M_1, M_2 \subset X$ and any pair $A, B \in \mathcal{L}(E)$ of local-type operators.

It is readily checked that $L_p(X, d\mu) \in k$ for all p, $1 \le p \le \infty$. In these spaces the estimate

$$\|P_{M_1} A P_{M_1} + P_{M_2} B P_{M_2}\| \le max(\|A\|, \|B\|) \, ,$$

which implies (6.2), holds for any pair of operators A, B (not necessarily of local type).

In [71] it is shown that the following estimate holds in $L_p(X)$ for any operator A of local type:

$$|A| \le (r + 1) \sup_{x \in X} q(x, A) \, ; \tag{6.3}$$

here r denotes the dimension of the topological space X (in [71] only finite-dimensional spaces X are considered). It turns out that the following result is valid.

THEOREM 6.1. *Let* $E \in k$. *Then the equality*

$$|A| = \sup_{x \in X} q(x, A) \tag{6.4}$$

holds for every operator $A \in \mathcal{L}(E)$ of local type.

To prove this theorem we need two auxiliary facts. Set

$$\psi(A) = \sup_{x \in X} q(x, A) \; . \tag{6.5}$$

Fix a number $\varepsilon > 0$. Then for every point $x \in X$ there is a neighborhood $U = U(x)$ such that $|P_U A| < q(x) + \varepsilon \leq \psi(A) + \varepsilon$. We extract from the family $\{U(x)\}_{x \in X}$ a finite cover $U_1, \ldots, U_n$ of X. Let $V_1, \ldots, V_n$ $V_k \subset U_k$, $k = 1, \ldots, n$, be closed disjoint sets, and put $M = \bigcup_{k=1}^{n} V_k$.

LEMMA 6.1. *The following estimate holds:*

$$|P_M A| \leq \psi(A) + \varepsilon \; . \tag{6.6}$$

PROOF. Let $\tilde{U}_k$, $V_k \subset \tilde{U}_k \subset U_k$, be open disjoint sets. We may assume that $\tilde{U}_k \cap M = V_k$. Then, in view of (6.2),

$$|P_M A| = \left| \sum_{k=1}^{n} P_M P_{\tilde{U}_k} A \right| = \left| \sum_{k=1}^{n} P_M P_{\tilde{U}_k} A(P_{\tilde{U}_k} + P_{X \setminus \tilde{U}_k}) \right|$$

$$= \left| P_M \sum_{k=1}^{n} P_{\tilde{U}_k} A P_{\tilde{U}_k} \right| \leq \left| \sum_{k=1}^{n} P_{\tilde{U}_k} A P_{\tilde{U}_k} \right|$$

$$\leq \sup_k |P_{U_k} A| \leq \psi(A) + \varepsilon \; ,$$

as claimed.

$\square$

LEMMA 6.2. *Let $U_1, \ldots, U_n$ be an open covering of X and let m, $m > n$ be a fixed positive integer. Then for every integer k, $1 \leq k \leq m$, there exist closed disjoint sets $V_{k1}^m, \ldots, V_{kn}^m$, such that $V_{k\ell}^m \subset U_\ell$ and every point $x \in X$ belongs to at most n of the sets*

$$G_k^m = X \setminus \bigcup_{\ell=1}^{n} V_{k\ell}^m \; .$$

PROOF. Let $f_1, \ldots, f_n$, $f_\ell \in C(X)$, be a partition of unity subordinate to the covering $U_1, \ldots, U_n$. Set

$$V_{k\ell}^m = \left\{ f_\ell \geq \frac{k+1}{n(m+1)} \; ; f_{\ell+1} \leq \frac{k}{n(m+1)} \; ; \ldots ; f_n \leq \frac{k}{n(m+1)} \right\} \quad \text{if} \quad \ell < n ,$$

$$V_{kn}^m = \left\{ f_n \geq \frac{k+1}{n(m+1)} \right\} ,$$

and

$$W_k^m = \bigcup_{\ell=2}^n \left\{ \frac{k}{n(m+1)} < f_\ell < \frac{k+1}{n(m+1)} \right\} .$$

Since $f_\ell(x) = 0$ in $X \backslash U_\ell$, $V_{k\ell}^m \subset U_\ell$. If $p > q$, then $f_p(x) \geq \frac{k+1}{n(m+1)}$ for $x \in V_{kp}^m$, and $f_p(x) \leq \frac{k}{n(m+1)}$ for $x \in V_{kq}^m$; consequently, $V_{kp}^m \cap V_{kq}^m = \emptyset$. We show that $G_k^m \subset W_k^m$. Let $x \notin W_k^m$; then there are three possible cases:

1) $f_n(x) \geq \frac{k+1}{n(m+1)}$ and then $x \in V_{kn}^m$;

2) $f_n(x) \leq \frac{k}{n(m+1)}$, $\ldots , f_{\ell+1}(x) \leq \frac{k}{n(m+1)}$, $f_\ell(x) > \frac{k}{n(m+1)}$ $\quad (\ell \geq 2)$. Since $x \notin W_k^m$, $f_\ell(x) \geq \frac{k+1}{n(m+1)}$, and hence $x \in V_{k\ell}^m$;

3) $f_\ell(x) \leq \frac{k}{n(m+1)}$, $\ell = 2, \ldots, n$. In this case $f_1(x) = 1 - \Sigma_{\ell=2}^n f_\ell(x) \geq 1 - \frac{k(n-1)}{n(m+1)} \geq \frac{k+1}{n(m+1)}$, and hence $x \in V_{k1}^m$.

To complete the proof of the lemma it suffices to check that each point $x \in X$ belongs to less than n of the sets $W_1^m, W_2^m, \ldots, W_m^m$.

Suppose that x belongs to n distinct sets $W_{k_1}^m, \ldots, W_{k_n}^m$. Then there are indices $\ell_1, \ldots, \ell_n$, $2 \leq \ell_i \leq n$, such that

$$\frac{k_i}{n(m+1)} < f_{\ell_i}(x) < \frac{k_i+1}{n(m+1)} .$$

But two of the indices ℓ_i must coincide, which contradicts the assumption that $k_1, \ldots, k_n$ are distinct.

$\square$

PROOF OF THEOREM 6.1. Let

$$M_k^m = \bigcup_{\ell=1}^n V_{k\ell}^m \, ,$$

where $V_{k\ell}^m$ are the sets constructed in Lemma 6.2, and

$$h_m = \sum_{k=1}^m \chi_{M_k^m} \, ,$$

where $\chi_{M_k^m}$ is the indicator function of the set M_k^m. Since every point $x \in X$ belongs to at most n of the sets G_k^m $(= X \backslash M_k^m)$, it follows that $m - n \le h_m(x) \le m$. Consequently,

$$|(m-n)A| \le |h_m A| = \left| \sum_{k=1}^m P_{M_k^m} A \right|$$

$$\le \sum_{k=1}^m |P_{M_k^m} A| \le m(\psi(A) + \varepsilon) \, .$$

Therefore,

$$|A| \le \frac{m}{m-n}(\psi(A) + \varepsilon).$$

Since ε is independent of m, we conclude that $|A| \le \psi(A)$. The opposite inequality is obvious.

□

Now let A_x, $x \in X$, and A, be local-type operators belonging to $\mathcal{L}(E)$. The operator A is called the *envelope* of the family $\{A_x\}$ if

$$\inf_{U(x)} |P_{U(x)}(A - A_x)| = 0$$

for every $x \in X$ (see [71]).

In [71] (see also [89]) it is proved, under certain restrictions on the family $\{A_x\}$, that an envelope A exists in $L_p(X)$ (with $dim\ X = r < \infty$), and the following bound is established:

$$|A| \le (r+1) \sup_x |A_x| \, .$$

With the help of Theorem 6.1 the proof of the existence of an envelope can be carried over to the case $E \in k$, with no restrictions on the dimension of the space X. Moreover, the exact estimate

$$|A| \leq \sup |A_x| \tag{6.7}$$

is valid.

In fact, it is readily checked that $q(x, A) = q(x, A_x)$. By Theorem 6.1,

$$|A| = \sup_x q(x, A) = \sup \ q(x, A_x)$$

$$\leq \sup_x \sup_y \ q(y, A_x) = \sup_x \ |A_x| \ .$$

We list a number of properties of the characteristic $q(x, A)$, the proof of which is a routine verification.

6.1°. Let $x_1, x_2 \in X$, $\alpha : X \longrightarrow X$ a homeomorphism with $\alpha(x_1) = x_2$, and let $\mu \in \mathbb{C}$. Suppose that the operator V defined by the formula $(V\varphi)(x) = \mu\varphi(\alpha(x))$, is isometric and invertible. If $A \in \mathcal{L}(E)$ and $V^{-1}AV - A \in T(E)$, then $q(x_1, A) = q(x_2, A)$.

6.2°. Let $A \in \mathcal{L}(E)$ and $x_0 \in X$. Suppose that there exists a neighborhood U of x_0 and a homeomorphism $\alpha : X \longrightarrow X$ with the following properties:

- for every neighborhood V of x_0 there is an $N \in \mathbb{N}$ such that $\alpha^N(U) \subset V$;

- the operator V, defined by $(V\varphi)(t) = \mu\varphi(\alpha(t))$, is invertible and isometric for at least one value of $\mu \in \mathbb{C}$;

- $A - VAV^{-1} \in T(E)$.

Then $|P_W A| = q(x_0, A)$ for every neighborhood W of x_0 such that $W \subset U$.

6.3°. Let $A, B \in \mathcal{L}(E)$ and suppose A is an operator of local type. Then $q(x, AB) \leq q(x, A)q(x, B)$.

Remark 6.1. If A is not of local type then 6.3° is not necessarily true. For example, take $E = L_2(-1, 1)$, $(A\varphi)(t) = \varphi(-t)$, and $B = P_{[-1,0]}$. For this choice $q(1, AB) = 1$, whereas $q(1, B) = 0$.

Remark 6.2. For operators which are not of local type Theorem 6.1 may fail too. For example, consider the operator

$$(A\varphi)(t) = \begin{cases} \varphi(t) & \text{if } t > 0 \\ \varphi(-t) & \text{if } t < 0 , \end{cases} \tag{6.8}$$

acting in $L_2(X)$, where $X = [-2,-1] \cup [1,2]$. Here $|A| = \sqrt{2}$, whereas $q(x, A) = 1$ for all $x \in X$.

Remark 6.3. Set

$$\tilde{q}(x, A) = \inf_{U(x)} |P_U A P_U| . \tag{6.9}$$

Then it is readily checked that for operators A of local type $q(x, A) = \tilde{q}(x, A)$, and hence (under the assumptions of Theorem 6.1)

$$|A| = \sup_x \tilde{q}(x, A) . \tag{6.10}$$

If, however, A is not of local type, (6.10) is not necessarily valid, as the example of the operator A defined by formula (6.8) shows: for this case $|A| = \sqrt{2}$, whereas

$$\tilde{q}(x, A) = \begin{cases} 1 & \text{if } x > 0 \\ 0 & \text{if } x < 0 . \end{cases}$$

Following [71], we say that the operators A and B of local type are *equivalent at the point* x_0 (and write $A \overset{x_0}{\sim} B$) if $\inf |P_U(A - B)| = 0$, where the infimum is taken over all the neighborhoods of x_0. It is easily seen that if $A \overset{x_0}{\sim} B$, then $q(x_0, A) = q(x_0, B)$. It follows from Theorem 6.1 that if the operator A of local type is equivalent at each point $x \in X$ with a compact operator A_x, then A is itself compact. A consequence of 6.3° is

6.4°. Let A_i and B_i, $j = 1, 2$, be operators of local type. If $A_j \overset{x_0}{\sim} B_j$ for $j = 1, 2$, then $A_1 A_2 \overset{x_0}{\sim} B_1 B_2$ and $A_1 + A_2 \overset{x_0}{\sim} B_1 + B_2$.

Let $E_j \in k$, $j = 1, 2$, be Banach spaces of functions $f : X_j \longrightarrow \mathbb{C}$, and let $A_j \in \mathcal{L}(E_j)$ and $x_j \in X_j$. By analogy with [71], we say (for operators of local type) that

A_1 and A_2 are *locally quasi-equivalent at the pair of points* x_1, x_2 (and write $A_1 \overset{x_1}{\sim} \overset{x_2}{\sim} A_2$), if there exists a homeomorphism $\alpha : X_1 \longrightarrow X_2$ and a function $h \in C(X_1)$ such that $\alpha(x_1) = x_2$, the operator V from E_2 to E_1, defined by $(V\varphi)(x) = h(x)\varphi(\alpha(x))$, is invertible and isometric, and $V^{-1}A_1 V \overset{x_2}{\sim} A_2$.

6.5°. If $A_1 \overset{x_1}{\sim} \overset{x_2}{\sim} A_2$, then $q(x_1, A_1) = q(x_2, A_2)$.

We denote $$q_0(x, A) = \inf_{U(x)} \|P_U A\| ,$$

where the infimum is taken over all neighborhoods of the point x.

Recall that the space E is said to be *regular* if $\lim_{n \to \infty} \|\chi_{D_n} f\| = 0$ for every function $f \in E$ and any nonincreasing sequence $\{D_n\}$ of sets, the measure of which $\mu(D_n) \longrightarrow 0$ (here χ stands for the indicator function). The space $L_p(X)$ are regular for $1 \le p < \infty$, but not for $p = \infty$.

THEOREM 6.2. *In a regular space $E \in k$ the equality*

$$q_0(x, A) = q(x, A) , \quad x \in X , \tag{6.11}$$

holds for every operator A of local type.

PROOF. Fix $\varepsilon > 0$, $T \in \mathcal{T}(E)$, $x \in X$, and a neighborhood V of the point x. Since E is regular, $\|\chi_{D_n} T\| \longrightarrow 0$, and hence there is a neighborhood $U \subset V$ of x such that $\|P_U T\| > \varepsilon$. Then $\|P_V A + T\| \ge \|P_U (P_V A + T)\| \ge \|P_U A\| - \varepsilon \ge q_0(x, A) - \varepsilon$, which yields (6.11).

$\square$

COROLLARY 6.1. *In a regular space $E \in k$*

$$|A| = \sup_{x \in X} \inf_{U(x)} \|P_U A\| \tag{6.12}$$

for every operator A of local type.

$\square$

Let us examine a few examples:

Example 6.1. Let Γ_1 and Γ_2 be simple closed Lyapunov contours, $\alpha : \Gamma_1 \longrightarrow \Gamma_2$ a homeomorphism whose derivative α' satisfies Hölder's condition, $a, b \in \mathbb{C}$, $E_j = L_p(\Gamma_j)$, and $A_j = aI + bS_{\Gamma_j}$, $j = 1, 2$. Then $A_1 \overset{t_1}{\sim} \overset{t_2}{\sim} A_2$ whenever $t_1 \in \Gamma_1$ and $t_2 = \alpha(t_1) \in \Gamma_2$. Fix a point $t_1 \in \Gamma_1$. Then for each point $\tau \in \Gamma_2$ one can choose a homeomorphism α_τ such that $\alpha_\tau(t_1) = \tau$. By the foregoing discussion,

$$|A_1| \geq q(t_1, A_1) = q(\tau, A_2) = \sup_{\tau \in \Gamma_2} q(\tau, A_2) = |A_2| \, .$$

Since the spaces E_1 and E_2 are isometrically isomorphic, it follows that

$$|aI + bS_{\Gamma_1}|_{E_1} = |aI + bS_{\Gamma_2}|_{E_2} \, . \tag{6.13}$$

Example 6.2. Let Γ be a simple closed Lyapunov contour and $A = aI + bS_\Gamma$ (with $a, b \in C(\Gamma)$). We show that in $L_p(\Gamma)$

$$\begin{aligned} |aI + bS_\Gamma| &= \sup_{\tau \in \Gamma} |a(\tau)I + b(\tau)S_\Gamma| \\ &= \sup_{z \in \Gamma_0} |\tilde{a}(z) + \tilde{b}(z)S_0| \end{aligned} \tag{6.14}$$

in which $z = \alpha(\tau)$, $\alpha : \Gamma \longrightarrow \Gamma_0$ is the homeomorphism considered in the preceding example, and $\tilde{a}(z) = a(\alpha(z))$, $\tilde{b}(z) = b(\alpha(z))$.

By (6.13), it suffices to prove the first equality in (6.14). Let τ be an arbitrary point of Γ. Let A_τ denote the operator with constant coefficients $a(\tau)I + b(\tau)S_\Gamma$. We have seen that $q(t, A_\tau)$ does not depend on t; consequently

$$\begin{aligned} |A| &= \sup_{\tau \in \Gamma} q(\tau, A) = \sup_{\tau \in \Gamma} q(\tau, A_\tau) \\ &= \sup_{\tau \in \Gamma} \sup_{t \in \Gamma} q(t, A_\tau) = \sup_{\tau \in \Gamma} |A_\tau| \, , \end{aligned}$$

as claimed.

Example 6.3. Let Γ be a simple closed Lyapunov contour, and let $a, b \in C(\Gamma)$.

Then

$$|aI + bS_\Gamma|_{L_p(\Gamma)}$$

$$\geq \sup_{z\in\Gamma}\left[\left(|b(t)|^2\cot^2\frac{\pi}{p} + \left(\frac{|a(z)+b(z)| - |a(z)-b(z)|}{2}\right)^2\right)^{1/2} + \right.$$

$$\left. + \left(|b(z)|^2\cot^2\frac{\pi}{p} + \left(\frac{|a(z)+b(z)| + |a(z)-b(z)|}{2}\right)^2\right)^{1/2}\right] . \tag{6.15}$$

This follows from relations (6.14) and (4.2).

Example 6.4. Let Γ be a simple closed Lyapunov contour and let $\omega \in C(\Gamma\times\Gamma)$ be a function which satisfies Hölder's condition in one of its arguments. The operator A defined by

$$(A\varphi)(t) = \frac{1}{\pi}\int_\Gamma \frac{\omega(\tau,t)}{\tau-t}\varphi(\tau)d\tau \tag{6.16}$$

differs from the operator $\omega(t,t)S_\Gamma$ by a compact term. It follows from (6.14) and (4.18) that in $L_p(\Gamma)$

$$|A| = \nu(p)\max_{t\in\Gamma}|\omega(t,t)| , \tag{6.17}$$

where the number $\nu(p)$ is defined by (4.18).

Returning to our analysis of operators of local type, let $\rho_k : X \longrightarrow \mathbb{R}$, $k = 1,\ldots,n$, be measurable, nonnegative, a.e. finite and nonvanishing functions, and let $\rho(t) = \Pi_{k=1}^n\rho_k(t)$ and $\rho_0(t) \equiv 1$.

THEOREM 6.3. *Suppose that the operator A of local type is bounded in each of the spaces $L_p(X,\rho_k)$, $k=1,\ldots,n$. If for every point $\tau \in X$ one can find a neighborhood U of τ in which all the weights $\rho_k^{\pm1}(t)$ except, possibly for one, are continuous, then $A \in \mathcal{L}(L_p(X,\rho))$ and*

$$|A|_{p,\rho} \leq \max_k |A|_{p,\rho_k} . \tag{6.18}$$

PROOF. Let τ be an arbitrary point of X and suppose that in one of its neighborhoods all functions $\rho_k^{\pm 1}$ with $k \neq m$ are continuous. Then

$$q(\tau, A)_{p,\rho} = q(\tau, \rho^{1/p} A \rho^{-1/p})_p = q(\tau, \rho_m^{1/p} A \rho_m^{-1/p})$$
$$= q(\tau, A)_{p,\rho_m} \leq |A|_{p,\rho_m} \leq \max_{0 \leq k \leq n} |A|_{p,\rho_k} .$$

If now in a neighborhood of τ all the functions $\rho_k^{\pm 1}$ are continuous, then $q(\tau, A)_{p,\rho} = q(\tau, A)_p \leq |A|_p \leq \max_{0 \leq k \leq n} |A_p|_{\rho_k}$. To complete the proof, apply Theorem 6.1.

$\square$

THEOREM 6.4. *Suppose* Γ *is a simple closed Lyapunov contour and the operator* S_Γ *is bounded in* $L_p(X, \rho)$ *(with* $1 < p < \infty$*). Then for every choice of numbers* $a, b \in \mathbb{C}$

$$|aI + bS_\Gamma|_{p,\rho} \geq |aI + bS_\Gamma|_p . \tag{6.19}$$

[Concerning boundedness criteria for the operator S_Γ in $L_p(\Gamma, \rho)$ see Sec. 12.]

PROOF. Suppose first that $\Gamma = \Gamma_0$ and $S = S_{\Gamma_0}$. Let R_n denote the operator

$$R_n = \frac{1}{2}(t^{-n+1} S_0 t^{n-1} - t^n S_0 t^{-n})$$

which sends each function $\varphi \in L_p(\Gamma_0, \rho)$ into the partial sum of its Fourier series. Since $S_0 \in \mathcal{L}(L_0(\Gamma, \rho))$, it follows that R_n converges strongly to I. We define a sequence of operators $\{B_n\}$ by the rule $(B_n \varphi)(t) = t^n \varphi(t^{2n})$. If $A = aI + bS_0$, $a, b \in \mathbb{C}$, it is readily verified that $B_n A = A B_n$ and $R_n B_n = 0$. Moreover,

$$B_n A \varphi = (A + T R_n) B_n \varphi ,$$

for every operator $T \in \mathcal{T}(E)$ and every trigonometric polynomial φ, whence

$$\|B_n A \varphi\|_{p,\rho} \leq \|A + T R_n\|_{p,\rho} \|B_n \varphi\|_{p,\rho} . \tag{6.20}$$

We now use a well-known Lemma of Fejér [81, Vol.1], according to which

$$\lim_{n \to \infty} \int_{\Gamma_0} \varphi(t^n) \psi(t) |dt| = \frac{1}{2\pi} \int_{\Gamma_0} \varphi(t) |dt| \int_{\Gamma_0} \psi(t) |dt|$$

for every function $\psi \in L_1(\Gamma)$.

Since $S_0 \in \mathcal{L}(L_p(\Gamma_0, \rho))$, it follows (by Lemma 4.1) that $\rho \in L_1(\Gamma_0)$, and hence

$$\lim_{n \to \infty} \|B_n A\varphi\|_{p,\rho}^p = \frac{1}{2\pi} \|\rho\|_{L_1(\Gamma_0)} \|A\varphi\|_p^p \tag{6.21}$$

and

$$\lim_{n \to \infty} \|B_n\varphi\|_{p,\rho}^p = \frac{1}{2\pi} \|\rho\|_{L_1(\Gamma_0)} \|\varphi\|_p^p . \tag{6.22}$$

Now let T be an arbitrary compact operator. Then $\|TR_n - T\|_{p,\rho} \longrightarrow 0$. Letting $n \longrightarrow \infty$ in (6.20) and using (6.21) and (6.22) we get $\|A\varphi\|_p \leq \|A + T\|_{p,\rho}\|\varphi\|_p$. Consequently, $\|A\|_p \leq |A|_{p,\rho}$, and hence $|A|_p \leq |A|_{p,\rho}$.

We now turn to the general case of a Lyapunov contour Γ. Let $\alpha : \Gamma \longrightarrow \Gamma_0$ be a homeomorphism whose derivative α' satisfies Hölder's condition on Γ, and let $(V\varphi)(t) = |\alpha'(t)|^{1/p}\varphi(\alpha(t))$. Then V is an isometric mapping of $L_p(\Gamma)$ onto $L_p(\Gamma_0)$, and also of $L_p(\Gamma, \rho)$ onto $L_p(\Gamma_0, \rho_0)$, where $\rho_0(\alpha(t)) = \rho(t)$. Furthermore, the operator $S_0 - VS_\Gamma V^{-1}$ is compact (see, for example, [50(9)]). It follows that

$$|A|_{L_p(\Gamma, \rho)} = |A|_{L_p(\Gamma_0, \rho_0)} \geq |A|_{L_p(\Gamma_0)} = |A|_{L_p(\Gamma)} ,$$

as claimed.

$\square$

Notice that in the proof we derived the bound $\|A\|_{L_p(\Gamma_0)} \leq |A|_{L_p(\Gamma_0)}$. For $\rho(t) \equiv 1$ we obtain the equality $\|A\|_{L_p(\Gamma_0)} = |A|_{L_p(\Gamma_0)}$ that we already encountered (see (4.9)).

THEOREM 6.5. *Suppose Γ is a simple closed Lyapunov contour, the operator S_Γ is bounded in each space $L_p(\Gamma, \rho_k)$, $k = 1, \ldots, n$, and for every point $\tau \in \Gamma$ there is a neighborhood U in which all the weights $\rho_k^{\pm 1}$ except, possibly for one, are continuous. Let $\rho = \rho_1 \ldots \rho_n$. Then for every choice of numbers $a, b \in \mathbb{C}$*

$$\begin{aligned} |aI + bS_\Gamma|_{L_p(\Gamma, \rho)} &= \max_{1 \leq k \leq n} |aI + bS_\Gamma|_{L_p(\Gamma, \rho_k)} \\ &= \max_{z \in \Gamma} |a(z)I + b(z)S_\Gamma|_{L_p(\Gamma, \rho_z)} \end{aligned} \tag{6.23}$$

By Lemma 4.1, the conditions $-1 < \alpha_k < 1$ are necessary for the operator S_Γ to be bounded in $L_p(\Gamma, \rho)$. For $p \geq 2$ we set

$$\eta(p, \alpha) = \begin{cases} \cot \dfrac{\pi(1 + \alpha)}{p} & \text{if} \quad -1 < \alpha < -\frac{1}{2} \quad , \\[3mm] \cot \dfrac{\pi}{2p} & \text{if} \quad -\frac{1}{2} \leq \alpha \leq p - \frac{3}{2} \quad , \\[3mm] \cot \dfrac{\pi(p - 1 - \alpha)}{p} & \text{if} \quad p - \frac{3}{2} < \alpha < p - 1 \quad , \end{cases}$$

and for $p \geq 2$

$$\eta(p, \alpha) = \eta(q, \alpha(1 - q)) \, ,$$

where $p^{-1} + q^{-1} = 1$.

THEOREM 7.1. *Suppose that ρ is given by formula (7.1). Then*

$$|S_\Gamma|_{p,\rho} = \max_{1 \leq k \leq 2m} \eta(p, \alpha_k) \, . \tag{7.2}$$

The proof rests on a number of lemmas. We put $\Delta = [-1, 1]$.

LEMMA 7.1.

$$|S_\Gamma|_{p,\rho} = \max_j |S_{\Gamma_j}|_{p,\rho_j} \, ,$$

where

$$\rho_j(t) = |t - t_j|^{\alpha_j} |t - t_{j+m}|^{\alpha_{j+m}} \, .$$

PROOF. By Theorem 6.1,

$$|S_\Gamma|_{p,\rho} = \sup_k \sup_{\tau \in \Gamma_k} q(\tau, S_\Gamma)_{p,\rho}$$

$$= \sup_k \sup_{\tau \in \Gamma_k} q(\tau, S_{\Gamma_k})_{p,\rho_k}$$

$$= \sup_k |S_{\Gamma_k}|_{p,\rho}$$

□

LEMMA 7.2. *Let a and b denote the extremities of the open Lyapunov contour* Γ. *Then*

$$|S_\Gamma|_{p,|t-a|^\alpha|t-b|^\beta} = |S_\Gamma|_{p,|1-t|^\alpha|1+t|^\beta} .$$

PROOF. Find a homeomorphism $\alpha : [0,1] \longrightarrow \Gamma$ whose derivate α' satisfies Hölder's condition on $[0,1]$, and then use property 6.5° and Theorem 6.1.

□

LEMMA 7.3.   *Let φ be a continuously differentiable function on Δ which vanishes in neighborhoods of the points ± 1. Then the following identities hold (see [30]):*

$$\left. \begin{array}{c} \dfrac{1}{2\pi} \displaystyle\int_{-\pi}^{\pi} \dfrac{sgn\ \theta\varphi(\cos\theta)}{\tan\frac{\tau-\theta}{2}}d\theta = \dfrac{1}{\pi}\displaystyle\int_{-1}^{1}\dfrac{\varphi(x)}{x-y}dx \\[3mm] and \\[3mm] \dfrac{1}{2\pi}\displaystyle\int_{-\pi}^{\pi}\dfrac{\varphi(\cos\theta)}{\tan\frac{\tau-\theta}{2}}d\theta = \dfrac{1}{\pi}\displaystyle\int_{-1}^{1}\sqrt{\dfrac{1-y^2}{1-x^2}}\dfrac{\varphi(x)}{x-y}dx , \end{array} \right\} \qquad (7.3)$$

where $y \in \Delta$ and $\tau = arccos\ y$.

These identities are verified with the help of the substitution $x = \cos\theta$.

□

LEMMA 7.4. *Suppose $1 < p < \infty$, $-1 < \alpha < 1$, and $-1 < \beta < p-1$. Then*

$$|S_\Delta|_{p,(1-t)^\alpha(1+t)^\beta} \geq \max(\eta(p,\alpha), \eta(p,\beta)) . \qquad (7.4)$$

PROOF. Let $E = L_p(\Delta, (1-t)^\alpha(1+t)^\beta)$. We consider the operator

$$A = I + i\ tan(\pi/r)S_\Delta$$

in E. A criterion for A to be a Fredholm operator was obtained in [50(9)]. In particular, if the number r takes one of the values

$$p(1+\beta)^{-1}, \quad -p(1+\alpha)^{-1}, \quad p(p-1-\alpha)^{-1}, \quad \text{or} \quad -p(p-1-\beta)^{-1} ,$$

then $A \notin \Phi(E)$. This implies that $|tan(\pi/r)S_\Delta| \geq 1$, i.e.,

$$|S_\Delta|_E \geq \cot\frac{\pi}{r} \qquad (7.5)$$

for the indicated four values of r. Moreover,

$$\begin{aligned}
|S_\Delta|_E &= \sup_{t\in\Delta} q(t, S_\Delta)_E \leq q(\tau, S_\Delta)_{p,\rho} \\
&= q(\tau, S_\Delta)_p = q(\tau, S_\gamma)_p = |S_\gamma|_p \qquad (7.6) \\
&= \cot\frac{\pi}{2p}
\end{aligned}$$

where $\bar{p} = \max(p, p(p-1)^{-1})$. Inequality (7.4) follows from (7.5) and (7.6).

$\square$

LEMMA 7.5.   I.   *If $\alpha, \beta \in (-1, \frac{p}{2} - 1)$, then*

$$\|S_\Delta\|_E \leq \|S_0\|_{p,|t-1|^{2\alpha+1}|t+1|^{2\beta+1}} . \qquad (7.7)$$

II.   *If $\alpha, \beta \in (\frac{p}{2} - 1, p - 1)$, then*

$$\|S_\Delta\|_E \leq \|S_0\|_{p,|t-1|^{2\alpha+1-p}|t+1|^{2\beta+1-p}} . \qquad (7.8)$$

PROOF.   It is readily checked, with the help of the substitution $t = \cos\theta$ ($\theta \in [-\pi, \pi]$), that

$$\int_{-1}^{1} |\varphi(t)|^p (1-t)^\alpha (1+t)^\beta \, dt = 2^{-\alpha-\beta-1} \int_{-\pi}^{\pi} |\varphi(\cos\theta)^p| \, |e^{i\theta} - 1|^{2\alpha+1}|e^{i\theta} + 1|^{2\beta+1} d\theta .$$

Let C be the operator defined by formula (4.18). Using the first of identities (7.3), we get

$$\|S_\Delta \varphi\|_E^p = 2^{-\alpha-\beta-1} \int_{-\pi}^{\pi} |Cg|^p |e^{i\theta} - 1|^{2\alpha+1}|e^{i\theta} + 1|^{2\beta+1} d\theta ,$$

where $g(\theta) = sgn\, \theta\, \varphi(\theta)$. Consequently,

$$\|S_\Delta\|_E \leq \|C\|_{p,|e^{i\theta}-1|^{2\alpha+1}|e^{i\theta}+1|^{2\beta+1}} .$$

In Sec.4 we remarked that if ξ designates the isometric mapping $(\xi\varphi)(x) = \varphi(e^{ix})$, then $\xi C \xi^{-1} = i(K - S_0)$, where K is a rank-one operator. Let

$$\rho(t) = |t - 1|^{2\alpha+1}|t + 1|^{2\beta+1} \quad (t \in \Gamma_0)$$

and

$$r(\theta) = \rho(exp\ i\theta) \quad (-\pi \leq \theta \leq \pi).$$

Then

$$\|C\|_{p,r} = \|S_0 - K\|_{p,\rho}.$$

Consider the operator u acting according to the rule $(u\varphi)(t) = \varphi(1/t)$, $t \in \Gamma_0$. It is readily checked that

$$uS_0u = 2K - S_0 \quad \text{and} \quad \|u\varphi\| = \|\varphi\|.$$

Consequently,

$$\|S_0\|_{p,\rho} = \|S_0 - 2K\|_{p,\rho},$$

whence

$$\|C\|_{p,r} = \|S_0 - K\|_{p,\rho} \leq \frac{1}{2}(\|S_0\|_{p,\rho} + \|S_0 - 2K\|_{p,\rho}) = \|S_0\|_{p,\rho},$$

which proves inequality (7.7).

Inequality (7.8) is established analogously, using the second of equalities (7.3) instead of the first.

$\square$

PROOF OF THEOREM 7.1. By Lemmas 7.1, 7.2, 7.3 and 7.4, it suffices to show that

$$|S_\Delta|_{p,(1-t)^\alpha(1+t)^\beta} \leq \max(\eta(p,\alpha), \eta(p,\beta)). \tag{7.9}$$

Assume, for the sake of definiteness, that $2 \leq p < \infty$. It follows from Theorem 5.1 and Lemma 7.5 that for $\alpha \in (-1, -\frac{1}{2}]$

$$\|S_\Delta\|_{p,(t-1)^\alpha(t+1)^{-1/2}} \leq \|S_0\|_{p,|1-t|^{2\alpha+1}} = cot\frac{\pi(1 + \alpha)}{p}, \tag{7.10}$$

whereas for $\alpha \in [p - \frac{3}{2}, p - 1)$

$$\|S_\Delta\|_{p,(1-t)^\alpha(1+t)^{1/2}} \leq \|S_0\|_{p,|1-t|^{2\alpha+1-p}} = \cot\frac{\pi(p-1-\alpha)}{p} \ . \tag{7.11}$$

In particular

$$\|S_\Delta\|_{p,(1-t)^{-1/2}(1+t)^{-1/2}} \leq \cot\frac{\pi}{2p}$$

and

$$\|S_\Delta\|_{p,(1-t)^{p-3/2}(1+t)^{p-1}} \leq \cot\frac{\pi}{2p} \ .$$

According to Theorem 4.9, for every $\alpha \in (-\frac{1}{2}, p - \frac{3}{2})$ there is a γ such that

$$\|S_\Delta\|_{p,(1-t)^\alpha(1+t)^\gamma} \leq \cot\frac{\pi}{2p} \ . \tag{7.12}$$

It follows from inequalities (7.10), (7.11), and (7.12) that for every $\alpha \in (-1, p-1)$ there is a γ such that

$$\|S_\Delta\|_{p,(1-t)^\alpha(1+t)^\gamma} \leq \eta(p, \alpha) \ .$$

This inequality remains valid on replacing Δ by the segment $\Delta_1 = [0, 1]$. But the operator $(1+t)^{\gamma/p}S_{\Delta_1}(1+t)^{-\gamma/p} - S_{\Delta_1}$ is compact, and hence

$$
\begin{aligned}
|S_\Delta|_{p,(1-t)^\alpha} &= |S_{\Delta_1}|_{p,(1-t)^\alpha} \\
&= |(1+t)^{\gamma/p}S_{\Delta_1}(1+t)^{-\gamma/p}|_{p,(1-t)^\alpha} \\
&= |S_{\Delta_1}|_{p,(1-t)^\alpha(1+t)^\gamma} \leq \eta(p, \alpha)
\end{aligned}
$$

(in establishing the first equality we used Lemma 7.2). In a similar manner one proves that

$$|S_\Delta|_{p,(1+t)^\beta} \leq \eta(p, \beta) \ .$$

Inequality (7.9) is now obtained upon applying Theorem 6.3.

$\square$

8. ANALOGUES FOR SPACES OF VECTOR-VALUED FUNCTIONS

Let Γ be a simple closed Lyapunov contour, ρ a weight of the form $\rho(t) = \Pi_{k=1}^{n}|t - t_k|^{\alpha_k}$ $(t_1, \ldots, t_n \in \Gamma)$, and E_0 a separable Hilbert space. We let $L_p(\Gamma, E_0, \rho)$, $1 < p < \infty$, denote the Banach space of weakly-measurable vector-valued functions $f : \Gamma \rangle E_0$ (see [73]) for which the norm

$$\|f\|_E = \left(\int_{\Gamma} \|f(t)\|_{E_0}^p \rho(t)|dt| \right)^{1/p}$$

is finite. As in the scalar case $(E_0 = \mathbb{C})$, if $1 < p < \infty$ and $-1 < \alpha_k < p-1$ $(k = 1, \ldots, n)$, then the operator S_Γ, defined by the formula

$$(S_\Gamma \varphi)(t) = \frac{1}{\pi i} \int_{\Gamma} \frac{\varphi(\tau) d\tau}{\tau - t} \quad (t \in \Gamma),$$

is bounded in $L_p(\Gamma, E_0, \rho)$ (see, e.g., [81, Vol. II, Lemma 5.18]).

THEOREM 8.1. *Suppose* $1 < p < \infty$, $-1 < \alpha < p - 1$, *and* $\rho_0(t) = |t - t_0|^\alpha$ *with* $t_0 \in \Gamma_0$. *Then*

$$\|S_\Gamma\|_{L_p(\Gamma_0, E_0, \rho_0)} = \nu(p, \alpha) .$$

PROOF. Consider the singular integral operator

$$(M\varphi)(x) = \frac{1}{2\pi} \int_{-\pi}^{\pi} \frac{\varphi(y) dy}{\sin \frac{y-x}{2}} . \qquad (8.1)$$

It is readily verified that

$$(S_0 f)(t) = \frac{e^{-ix/2}}{2\pi} \int_{-\pi}^{\pi} \frac{f(e^{iy}) e^{iy/2}}{\sin \frac{y-x}{2}} dy \qquad (8.2)$$

where $t = e^{ix}$. We put $\Delta = [-\pi, \pi]$ and $r(x) = |e^{ix} - 1|^\alpha$, $x \in \Delta$. Then

$$\|M\|_{L_p(\Delta, r)} = \|S_0\|_{p, \rho_0} .$$

Let $\{e_j\}$ be an orthonormal basis in E_0. Every function $f \in L_p(\Delta, E_0, r)$ can be uniquely written as $f(t) = \Sigma_j f_j(t) e_j$, where $f_j \in L_p(\Delta, r)$. Moreover,

$$\|f\|_{L_p(\Delta, E_0, r)} = \int_{-\pi}^{\pi} \left(\sum_j |f_j(t)|^2 \right)^{p/2} r(t) dt .$$

We shall use the same letter M to denote the operator acting in $L_p(\Delta, E_0, r)$ according to the rule

$$(Mf)(t) = \sum_j (Mf_j)(t) e_j .$$

Notice that

$$\|Mf\|^p_{L_p(\Delta, E_0, r)} = \int_{-\pi}^{\pi} \left(\sum_j |(Mf_j)(t)|^2 \right)^{p/2} r(t) dt .$$

Since M takes real-valued functions again into such functions,

$$\int_{-\pi}^{\pi} \left(\sum_j |(Mf_j)(t)|^2 \right)^{p/2} r(t) dt$$

$$\leq \|M\|^p_{L_p(\Delta, r)} \int_{-\pi}^{\pi} \left(\sum_j |f_j(t)|^2 \right)^{p/2} r(t) dt$$

(see the Zygmund-Marcinkiewicz lemma in [81, Vol. II]). Hence

$$\|Mf\|_{L_p(\Delta, E_0, r)} \leq \|M\|_{L_p(\Delta, r)} \|f\|_{L_p(\Delta, E_0, r)}$$

for every function $f \in L_p(\Delta, E_0, r)$. Taking $f(x) = exp(-ix/2) g(e^{ix})$, $x \in \Delta$, we obtain

$$\|S_0 g\|_{L_p(\Gamma_0, E_0, \rho_0)} \leq \|M\|_{L_p(\Delta, r)} \|g\|_{L_p(\Gamma_0, E_0, \rho_0)} ,$$

whence

$$\|S_0\|_{L_p(\Gamma_0, E_0, \rho_0)} \leq \|S_0\|_{p, \rho_0} = \nu(p, \alpha) .$$

$\square$

The converse inequality is obvious.

__Remark 8.1.__ If $dim\ E_0 = \infty$, then Corollary 6.2 fails.

In fact, let $f \in L_p(\Gamma, \rho)$. Then the sequence $f_n(t) = f(t)e_n$ converges weakly to zero in $E = L_p(\Gamma, E_0, \rho)$. Therefore,

$$\lim_{n)\infty} \frac{\|(S_0 + T)f_n\|}{\|f_n\|} = \frac{\|S_0 f\|_{L_p(\Gamma,\rho)}}{\|f\|_{L_p(\Gamma,\rho)}}$$

for all operators $T \in \mathcal{T}(E)$, i.e., $|S_0|_E \geq \|S_0\|_{L_p(\Gamma,\rho)}$. It follows from Theorem 5.3 that the norm $\|S_0\|_{p,\rho}$ depends, generally speaking, on the disposition of the points $t_1, \ldots, t_n$ on Γ.

However, the following result is valid.

THEOREM 8.2. _Let dim $E_0 < \infty$. Then_

$$|S_0|_{L_p(\Gamma, E_0, \rho)} = \max_{1 \leq k \leq n} \nu(p, \alpha_k) \ .$$

PROOF. The inequality

$$|S_0|_{L_p(\Gamma, E_0, \rho)} \geq |S_0|_{L_p(\Gamma_0, \rho)} = \max\ \nu(p, \alpha_k)$$

is obvious. Let us establish the opposite inequality. By analogy with the scalar case (see (6.13)) one can show that the norm $|S_\Gamma|_{L_p(\Gamma, E_0, \rho)}$ does not depend on the form of the closed Lyapunov contour Γ, so that we may assume that $\Gamma = \Gamma_0$.

It follows from equality (8.2) that

$$|S_0|_{L_p(\Gamma_0, E_0, \rho)} = |M|_{L_p(\Delta, E_0, r)} \ .$$

For each $T \in \mathcal{T}(L_p(\Delta, r))$ we define an operator $\tilde{T} \in \mathcal{T}(L_p(\Delta, E_0, r))$ by the formula

$$(\tilde{T}f)(t) = \sum_j (Tf_j)(t)e_j \ .$$

We let $L_p^{\mathbb{R}}(\Gamma, \rho)$ denote the L_p-space of real-valued functions. Suppose that the operator T takes real-valued functions again into such functions. By the Zygmund-Marcinkiewicz lemma [81, Vol. II]

$$\|M + \tilde{T}\|_{L_p(\Delta, E_0, r)} = \|M + T\|_{L_p(\Delta, r)} = \|M + T\|_{L_p^{\mathbb{R}}(\Delta, r)} \, ,$$

whence

$$|M|_{L_p(\Delta, E_0, r)} \leq \inf_{T \in \mathcal{T}} \|M + T\|_{L_p^{\mathbb{R}}(\Delta, r)} = |M|_{L_p^{\mathbb{R}}(\Delta, r)} \, .$$

Since M is an operator of local type, Theorem 6.3 yields

$$|M|_{L_p^{\mathbb{R}}(\Delta, r)} \leq \max_{1 \leq k \leq n} |M|_{L_p^{\mathbb{R}}(\Delta, r_k)} \, ,$$

where $r_k(x) = |e^{ix} - t_k|^{\alpha_k}$, $x \in \Delta$. But

$$\|M\|_{L_p^{\mathbb{R}}(\Delta, r_k)} \leq \|M\|_{L_p(\Delta, r_k)} = \|S_0\|_{p, \rho_k} = \nu(p, \alpha_k) \, .$$

Therefore

$$|S_0|_{L_p(\Gamma, E_0, \rho)} = |M|_{L_p(\Delta, E_0, r)} \leq \max_k \nu(p, \alpha_k) \, .$$

□

CHAPTER III

SINGULAR INTEGRAL OPERATORS WITH
MATRIX COEFFICIENTS

In this chapter we establish criteria for singular integral operators with piecewise-continuous matrix coefficients to be Fredholm operators, as well as conditions under which singular integral operators with bounded coefficients are Fredholm operators.

9. SOME INFORMATION ON SINGULAR INTEGRAL OPERATORS

We give the properties of operators $aP+bQ$ (with $P = \frac{1}{2}(I+S_\Gamma)$ and $Q = I-P$) that will be needed below. The proofs of various assertions can be found, for example, in $[18, 50(9), 60, 61]$.

Let Γ be a closed oriented Lyapunov contour which divides the complex plane $\mathbb{C}$ into two domains: $D^+(\ni 0)$ and $D^-(\ni \infty)$. We let $PC(\Gamma)$ denote the set of functions $a \in L_\infty(\Gamma)$ which have a finite number of first-kind discontinuities and no other discontinuities.

Let $t_1, \ldots, t_n$ be distinct points on Γ, let $p, \beta_1, \ldots, \beta_n$ be numbers such that $1 < p < \infty$ and $-1 < \beta_k < p-1$, and set $\rho(t) = \Pi_{k=1}^n |t - t_k|^{\beta_k}$. We put

$$\delta(t) = \begin{cases} 2\pi/p & \text{if} \quad t \in \Gamma \backslash \{t_1, \ldots, t_n\} \,, \\ 2\pi(1 + \beta_k)/p & \text{if} \quad t = t_k \,, \quad k = 1, \ldots, n \,, \end{cases}$$

$$\theta(t) = \pi - \delta(t) \,,$$

and

$$f(t,\mu) = \begin{cases} \dfrac{\sin(\theta\mu)}{\sin\theta}\,exp(i\theta(\mu-1)) & \text{if} \quad \delta(t) \neq \pi \,, \\[3mm] \mu & \text{if} \quad \delta(t) = \pi \,. \end{cases} \tag{9.1}$$

With each $a \in PC(\Gamma)$ we associate the function

$$a_{p,\rho}(t,\mu) = a(t+0)f(t,\mu) + a(t-0)(1-f(t,\mu)) \,, \tag{9.2}$$

$$t \in \Gamma \,, \quad 0 \le \mu \le 1 \,.$$

The range $W_{p,\rho}$ of $a_{p,\rho}$ is the union of the range of a and a number of arcs of circles and segments joining the points $a_k(t+0)$ and $a(t_k-0)$.

The function a is called $p, \rho\text{-}nonsingular$ if $a_{p,\rho}(t,\mu) \neq 0$ for all $(t,\mu) \in \Gamma \times [0,1]$. The number of times that the curve $W_{p,\rho}$ winds around the point $\lambda = 0$ in the positive direction will be denoted by $ind_{p,\rho}a$ or $ind\ a_{p,\rho}(t,\mu)$.

THEOREM 9.1.   *Let $a,b \in PC(\Gamma)$. In order for $A = aP + bQ$ (where $P = \frac{1}{2}(I + S_\Gamma)$ and $Q = I - P$) to be a Φ_+- or Φ_--operator in the space $E = L_p(\Gamma, \rho)$ it is necessary and sufficient that*

$$a(t+0)b(t-0)f(t,\mu) + a(t-0)b(t+0)(1-f(t,\mu)) \neq 0$$

for all $(t,\mu) \in \Gamma \times [0,1]$.

If this condition is satisfied, then $A \in \Phi(E)$ and

$$Ind\ A = -ind_{p,\rho}(a/b) \,.$$

$\square$

THEOREM 9.2.   *Let $a,b \in L_\infty(\Gamma)$. If $A = aP + bQ$ is a Φ_+- or Φ_--operator, then it is invertible from at least one side.*

$\square$

THEOREM 9.3. *Let* $a, b \in L_\infty^{n \times n}(\Gamma)$. *If* $A = aP + bQ$ *is a* Φ_+- *or* Φ_--*operator then* ess inf $|\det a(t)| > 0$ *and* ess inf $|\det b(t)| > 0$.

$\square$

We let $L_p^+(\Gamma, \rho)$ $(\overset{\circ}{L}_p^-(\Gamma, \rho))$ denote the set of functions $a \in L_p(\Gamma, \rho)$ which admit the Cauchy integral representation

$$a(t) = \lim_{z \to t} \frac{\pm 1}{2\pi i} \int_\Gamma \frac{a(\tau) d\tau}{\tau - z} \qquad (\text{a.e. } t \in \Gamma) , \tag{9.3}$$

where $z \in D^\pm$ and tends to t along any path which is not tangent to Γ.

If, in particular, the operator S_Γ is bounded in $L_p(\Gamma, \rho)$, then (see [50(9)])

1) $a \in L_1(\Gamma)$. In fact, it follows from Lemma 4.1 that $\rho^{-1/p} \in L_q(\Gamma)$. Since, in addition, $a\rho^{1/p} \in L_p(\Gamma)$, $a \in L_1(\Gamma)$.

2) $a \in L_p^+(\Gamma, \rho) \Longleftrightarrow (Pa = a; \ a \in L_p(\Gamma, \rho))$.

3) $a \in \overset{\circ}{L}_p^-(\Gamma, \rho) \Longleftrightarrow (Qa = a; \ a \in L_p(\Gamma, \rho))$.

This follows from (9.3) and the well-known Sokhotzki formulas

$$\lim_{\substack{z \to t \\ |z| \gtrless 1}} \frac{1}{2\pi i} \int_\Gamma \frac{a(\tau) d\tau}{\tau - z} = \frac{1}{2}((S_\Gamma a)(t) \pm a(t)) .$$

We set

$$L_p^-(\Gamma, \rho) = \{\varphi + \alpha : \varphi \in \overset{\circ}{L}_p^-(\Gamma, \rho), \ \alpha \in \mathbb{C}\} .$$

<u>Definition 9.1.</u> We say that the function $a \in L_\infty(\Gamma)$ *admits a* p, ρ-*factorization* (and write $a \in Fact(p, \rho)$) if it can be expressed as $a = a_- t^\kappa a_+$, where $\kappa \in \mathbb{Z}$, $a_+^{\pm 1} \in L_1^+(\Gamma)$, and $a_-^{\pm 1} \in L_1^-(\Gamma)$, and if the operator $a_+^{-1} S_\Gamma a_+ I \in \mathcal{L}(L_p(\Gamma, \rho))$.

We note that here we deviate from the traditional requirements that

$$\begin{aligned} a_+ &\in L_q^+(\Gamma, \rho^{1-q}) , & a_+^{-1} &\in L_p^+(\Gamma, \rho) , \\ a_- &\in L_p^-(\Gamma, \rho) , & a_-^{-1} &\in L_q^-(\Gamma, \rho^{1-q}) . \end{aligned} \tag{9.4}$$

However, it follows from the definition of factorization that the operator S is bounded in the spaces $L_p(\Gamma, |a_+|^{-p}\rho)$ and $L_q(\Gamma, |a_+|^q \rho^{1-q}) = L_p^*(\Gamma, |a_+|^{-p}\rho)$. By Lemma 4.1,

$$|a_+|^{-p}\rho \in L_1(\Gamma), \qquad |a_+|^p \rho^{-1} \in L_{q-1}(\Gamma),$$
$$|a_+|^q \rho^{1-q} \in L_1(\Gamma), \quad \text{and} \quad |a_+|^{-q}\rho^{q-1} \in L_{p-1}(\Gamma).$$

Since $|a_-| \leq const\, |a_+|^{-1}$, this implies that $a_+ \in L_q(\Gamma, \rho^{1-q})$, $a_+^{-1} \in L_p(\Gamma, \rho)$, $a_- \in L_p(\Gamma, \rho)$, and $a_-^{-1} \in L_q(\Gamma, \rho^{1-q})$. Also taking into account that $a_+^{\pm 1} \in L_1^+(\Gamma)$ and $a_-^{\pm 1} \in L_1^-(\Gamma)$, we obtain (9.4).

THEOREM 9.4.   *Let $a \in L_\infty(\Gamma)$ and $A = aP + Q$. Then*

$$A \in \Phi(L_p(\Gamma, \rho)) \Longleftrightarrow a \in Fact(p, \rho).$$

If $a \in Fact(p, \rho)$, then

$$Ind\, A = -\kappa.$$

$\square$

THEOREM 9.5.   *Let α be a conformal mapping of the domain D, bounded by the contour Γ, into the unit disc $D_0 = \{\varsigma \in \mathbb{C}\,:\, |\varsigma| < 1\}$, and let $(V\varphi)(t) = \varphi(\alpha(t))$. Then the operator $V S_0 V^{-1} - S_\Gamma$ in $L_p(\Gamma, \rho)$ is compact.*

$\square$

THEOREM 9.6.   *Let $a \in GL_\infty(\Gamma)$ and $a_0(t) = a(t)/|a(t)|$. Then there exist invertible operators B_1 and B_2 such that $B_1(aP + Q)B_2 = a_0 P + Q$.*

$\square$

THEOREM 9.7.   *Let $a \in GL_\infty(\Gamma)$ and suppose that $a(t) > 0$ for a.e. $t \in \Gamma$. Then the function a admits a factorization $a = a_- a_+$, where*

$$a_+ = exp(P\ell n\, a) \in GL_\infty^+ \quad \text{and} \quad a_- = exp(Q\ell n\, a) \in GL_\infty^-.$$

$\square$

10. SINGULAR INTEGRAL OPERATORS WITH PIECEWISE-CONTINUOUS MATRIX COEFFICIENTS

Let Γ be a simple closed Lyapunov contour in the complex plane $\mathbb{C}$. We denote by $PC^{n \times n}(\Gamma)$ the set of $n \times n$ matrix-valued functions whose entries belong to $PC(\Gamma)$. Let $1 < p < \infty$, $-1 < \beta_k < p - 1$ $(k = 1, \ldots, n)$, $t_1, \ldots, t_n \in \Gamma$, and $\rho(t) = \Pi_{k=1}^{n} |t - t_k|^{\beta_k}$.

We say that the matrix-valued function a $(\in PC^{n \times n}(\Gamma))$ is p, ρ-nonsingular if

$$det(f(t, \mu)a(t + 0) + (1 - f(t, \mu))a(t - 0)) \neq 0 \qquad (10.1)$$

for all $(t, \mu) \in \Gamma \times [0, 1]$, where the function $f(t, \mu)$ is defined by formula (9.1). If $\rho(t) \equiv 1$ we simply say that a is p-nonsingular.

We orient the closed continuous curve

$$W_{p,\rho}(a) = det(f(t, \mu)a(t + 0) + (1 - f(t, \mu))a(t - 0)) \qquad (10.2)$$

in such a way that at the points of continuity of a the motion on this curve corresponds to the motion of the point t on Γ in the positive direction, while the motion along the supplementary arcs corresponds to the variation of μ from zero to one. The number of times that the curve $W_{p,\rho}(a)$ winds around the point $\lambda = 0$ will be denoted by $ind\, W_{p,\rho}(a)$ or, alternatively, by $ind\, det\, a_{p,\rho}(t, \mu)$.

To simplify notation, if $A \in \mathcal{L}(E)$ we shall denote the matrix diagonal operator $diag(A_1 A, \ldots, A)$ by the same letter A.

THEOREM 10.1. *Let Γ be a simple closed Lyapunov contour and let $a, b \in PC^{n \times n}(\Gamma)$. Then $A = aP + bQ$ is a Φ-operator in $L_p^n(\Gamma, \rho)$ if and only if the following conditions are satisfied:*

1) $det\, b(\pm 0) \neq 0$ for all $t \in \Gamma$;

2)   the matrix $c = b^{-1}a$ is p, ρ-nonsingular.

If conditions 1) and 2) are satisfied, then

$$Ind \ A = -ind \ W_{p,\rho}(c) \ . \tag{10.3}$$

First we prove two lemmas.

LEMMA 10.1. *Let* $h, g \in C^{n \times n}(\Gamma)$, $a \in L_\infty^{n \times n}(\Gamma)$, *and* $det(hg)(t) \neq 0$ *for all* $t \in \Gamma$. *Then*

$$aP + Q \in \Phi(E) \Longleftrightarrow hagP + Q \in \Phi(E) \ ,$$

where $E = L_p^n(\Gamma, \rho)$.

PROOF. Since $fP - Pf \in T(E)$ for every matrix-valued function $f \in C^{n \times n}(\Gamma)$,

$$hag \ P + Q = (aP + Q)(gP + h^{-1}Q) + T \ ,$$

with $T \in T(E)$. Moreover, the operator $g^{-1}P + hQ$ is a regularizer for $gP + h^{-1}Q$, and hence $gP + h^{-1}Q \in \Phi(E)$, which proves the assertion of the lemma.

$\square$

LEMMA 10.2. *Every matrix-valued function* $a \in PC^{n \times n}(\Gamma)$ *such that* $det \ a(t \pm 0) \neq 0$ *for all* $t \in \Gamma$ *can be expressed as*

$$a = hxg \tag{10.4}$$

where $h, g \in GC^{n \times n}(\Gamma)$ *and* x *is an upper triangular matrix-valued function which belongs to* $PC^{n \times n}(\Gamma)$.

PROOF. Let $\tau_1, \ldots, \tau_m$ be the points of discontinuity of the matrix a. Choose nonsingular constant matrices $g_1, \ldots, g_m$ such that the matrices $g_k a(\tau_k - 0)^{-1} a(\tau_k + 0) g_k^{-1}$, $k = 1, \ldots, m$, are upper triangular. Let g be a continuous nonsingular matrix-valued function on Γ such that $g(\tau_k) = g_k$. Now, for $x(t)$ we take any nonsingular upper-triangular matrix-valued function which is continuous at all points $t \in \Gamma \backslash \{\tau_1, \ldots, \tau_m\}$ and satisfies the conditions $x(\tau_k + 0) = g_k a(\tau_k - 0)^{-1} a(\tau_k + 0) g_k^{-1}$. It is readily verified that for $k = 1, \ldots, m$

$$a(\tau_k + 0) g_k^{-1} x(\tau_k + 0)^{-1} = a(\tau_k - 0) g_k^{-1} x_k(\tau_k - 0)^{-1} \ .$$

This guarantees that the matrix-valued function $h = ag^{-1}x^{-1}$ is continuous on Γ, which completes the proof of the lemma.

$$\square$$

PROOF OF THEOREM 10.1.   Suppose that $det\ b(t \pm 0) \neq 0$ for all $t \in \Gamma$. The matrix $c = b^{-1}a$ can be represented in the form (10.4): $c = hxg$. Since c, h, and g are p, ρ-nonsingular matrix-valued functions and since h and g are continuous, x is also p, ρ-nonsingular. Next, since x is a triangular matrix, its diagonal entries x_{mm} are p, ρ-nonsingular functions. It follows from Theorem 9.1 that the diagonal entries of the triangular operator matrix $xP + Q$ are Φ-operators in $L_p(\Gamma, \rho)$. It now follows from Theorem 2.2 that $xP + Q$ is a Φ-operator in $L_p^n(\Gamma, \rho)$. In view of Lemma 10.1, the sufficiency of conditions 1) and 2) is thus established.

Let us prove the necessity of these conditions. Suppose that $A = aP + bQ$ is a Φ-operator. Then ess $\inf |det\ a(t)| > 0$ and ess $\inf |det\ b(t)| > 0$. This assertion is valid for operator A with arbitrary coefficients from $L_\infty^{n \times n}$ (see Sec.9). From this it follows, in particular, that $det\ a(t \pm 0) \neq 0$ and $det\ b(t \pm 0) \neq 0$ for every $t \in \Gamma$. Again, we express the matrix-valued function $c = b^{-1}a$ in the form (10.4): $c = hxg$. By Lemma 10.1, $X = xP + Q$ is a Φ-operator. Since the matrix operator $X = [X_{jk}]$ has triangular form, X_{11} is a Φ_--operator (see Theorem 2.2). But X_{11} is a singular integral operator with a piecewise-continuous coefficient. By Theorem 9.1, X_{11} is a Φ-operator. This implies (also by Theorem 2.2) that X_{22} is a Φ-operator. Therefore, upon applying Theorem 2.2 successively we can prove that all the diagonal entries X_{jj} of the matrix X are Φ-operators. It now follows from Theorem 9.1 that the diagonal entries of the triangular matrix x are p, ρ-nonsingular functions, and hence that

$$det(f(t, \mu)x(t + 0) + (1 - f(t, \mu))x(t - 0)) \neq 0$$

for all $(t, \mu) \in \Gamma \times [0, 1]$. Since

$$h, g \in C^{n \times n}(\Gamma) \quad \text{and} \quad det\ h(t)g(t) \neq 0\ ,$$

it follows that

$$det(f(t, \mu)c(t + 0) + (1 - f(t, \mu))c(t - 0)) \neq 0$$

for all $(t, \mu) \in \Gamma \times [0,1]$. This proves the necessity of conditions 1) and 2). It remains to verify the validity of equality (10.3). By (10.4),

$$Ind\ A = Ind(xP + Q) + Ind(Pg + Qh^{-1}) \ .$$

It is well known (see [60]) that for continuous matrix-valued functions h and g

$$Ind(Pg + Qh^{-1}) = -ind\ det(gh) \ .$$

By Theorem 3.3,
$$Ind(xP + Q) = \sum_m Ind(x_{mm}P + Q) \ .$$

Moreover,
$$Ind(x_{mm}P + Q) = -ind_{p,\rho} x_{mm}$$

(see Theorem 9.1), so that

$$Ind(xP + Q) = -\sum ind_{p,\rho} x_{mm} = -ind\ W_{p,\rho}(x) \ .$$

We finally obtain

$$Ind\ A = -ind\ det(gh) - ind\ W_{p,\rho}(x) = ind\ W_{p,\rho}(c) \ .$$

$\square$

We give one more variant of the criterion for a singular integral operator with piecewise-continuous matrix coefficients to be a Fredholm operator.

THEOREM 10.2.   Let $a, b \in PC^{n \times n}(\Gamma)$, let $\tau_1, \ldots, \tau_m$ be all the points of discontinuity of the matrices a and b, and let $\rho(t) = \Pi_{k=1}^r |t - t_k|^{\beta_k}$, where $r \geq m$ and $t_1 = \tau_1, \ldots, t_m = \tau_m$. Then $A = aP + bQ$ is a Φ-operator in $L_p^n(\Gamma, \rho)$ if and only if the following conditions are satisfied:

1) $det\ a(t \pm 0) \neq 0$ and $det\ b(t \pm 0) \neq 0$ for all $t \in \Gamma$;

2)   The eigenvalues λ_{jk} of the matrix $c(\tau_k + 0)^{-1} c(\tau_k - 0)$, where $c = b^{-1} a$, do not lie

on the ray

$$arg\ z = \frac{2\pi(1 + \beta_k)}{p}\ .$$

(10.5)

Suppose that conditions 1), 2), and

$$\frac{1 + \beta_k}{p} - 1 < \frac{\alpha_{jk}}{2\pi} < \frac{1 + \beta_k}{p}\ ,$$

(10.6)

where $\alpha_{jk} = arg\ \lambda_{jk}$, are satisfied. Then

$$Ind\ A = -\frac{1}{2\pi} \oint_\Gamma d(arg\ det\ c(t)) + \sum_k \sum_j \alpha_{jk}\ .$$

(10.7)

PROOF. The necessity of condition 1) follows, for example, from Theorem 10.1. Suppose now that 1) holds. In this case the condition

$$det(f(t, \mu)c(t + 0) + (1 - f(t, \mu))c(t - 0)) \neq 0 \qquad (t \in \Gamma)$$

that the matrix-valued function c be p, ρ-nonsingular is equivalent to the conditions

$$det\left(c(\tau_k + 0)^{-1}c(\tau_k - 0) - \frac{f(\tau_k, \mu)}{f(\tau_k, \mu) - 1}\right) \neq 0\ , \qquad k = 1, \ldots, m\ .$$

It is readily verified that for $0 \leq \mu \leq 1$ the set of values of the function $f(\tau_k, \mu)(f(\tau_k, \mu) - 1)^{-1}$ is just the ray (10.5). Therefore, conditions 1) and 2) of Theorem 10.1 and conditions 1) and 2) of Theorem 10.2 are equivalent. It remains to prove relation (10.7).

Suppose first that the matrix-valued function c is triangular. Then

$$\begin{aligned}
Ind\ A = Ind(cP + Q) &= \sum Ind(c_{jj}P + Q) \\
&= -\sum_j ind_{p,\rho}c_{jj} \\
&= -\frac{1}{2\pi} \sum_{j=1}^n \left(\oint_\Gamma d(arg\ c_{jj}(t)) - \sum_{k=1}^m arg\ c_{jj}(\tau_k + 0)^{-1}c_{jj}(\tau_k - 0)\right)\ ,
\end{aligned}$$

where

$$\frac{1 + \beta_k}{p} - 1 < \frac{1}{2\pi}arg\ c_{jj}(\tau_k + 0)^{-1}c_{jj}(\tau_k - 0) < \frac{1 + \beta_k}{p}\ .$$

But

$$\sum_{i=1}^{n} \oint_{\Gamma} d(arg \ c_{jj}(t)) = \oint_{\Gamma} d(arg \ det \ c)$$

and the numbers $c_{jj}(\tau_k + 0)^{-1} c_{jj}(\tau_k - 0)$ are the eigenvalues of the triangular matrix $c(\tau_k + 0)^{-1} c(\tau_k - 0)$. We have thus proved relation (10.7) for a triangular matrix c. In the general case we express c in the form (10.4): $c = hxg$, where h and g are continuous matrix-valued functions, $ind \ det \ h(t) = ind \ det \ g(t) = 1$, and x is triangular. We saw in the proof of Theorem 10.1 that $Ind \ A = Ind(xP + Q)$. Moreover,

$$c_0 = c(\tau_k + 0)^{-1} c(\tau_k - 0) = g(\tau_k)^{-1} x(\tau_k + 0)^{-1} x(\tau_k - 0) g(\tau_k)$$

$$= g(\tau_k)^{-1} x_0 g(\tau_k) \ ,$$

from which it follows that the eigenvalues of the matrices c_0 and x_0 coincide. This completes the proof of the theorem.

$\square$

In studying Banach algebras generated by singular integral operators it is convenient to write the Fredholmness criterion in the following form.

Let $a, b \in PC^{n \times n}(\Gamma)$ and $A = aP + bQ$. We consider the $2n \times 2n$ matrix-valued function

$$\mathcal{A}_{p,\rho}(t, \mu)$$

$$= \begin{bmatrix} f(t,\mu)a(t+0) + (1 - f(t,\mu))a(t-0) & h(t,\mu)(b(t+0) - b(t-0)) \\ h(t,\mu)(a(t+0) - a(t-0)) & (1 - f(t,\mu))b(t+0) + f(t,\mu)b(t-0) \end{bmatrix} ,$$

(10.8)

where $h(t,\mu) = \sqrt{f(t,\mu)(1 - f(t,\mu))}$.

THEOREM 10.3. *Let* $a, b \in PC^{n \times n}(\Gamma)$. *Then* $A = aP + bQ$ *is a* Φ-*operator in* $E = L_p^n(\Gamma, \rho)$ *if and only if*

$$det \ \mathcal{A}_{p,\rho}(t, \mu) \neq 0 \quad for \ all \quad (t, \mu) \in \Gamma \times [0, 1] \ .$$

(10.9)

PROOF. If $A \in \Phi(E)$, then

$$det \ b(t \pm 0)a(t \pm 0) \neq 0 \ .$$

Moreover, from (10.8) it also follows, upon setting $\mu = 0$ and $\mu = 1$, that

$$det\ b(t \pm 0)a(t \pm 0) \neq 0 \ .$$

Let us show that

$$det\ \mathcal{A}_{p,\rho}(t,\mu) = det\ b(t-0)b(t+0)det(b^{-1}a)(t,\mu) \ . \tag{10.10}$$

When $\mu = 0$ and $\mu = 1$, equality (10.10) is obvious. To prove it for the other values of μ we multiply the first (block) row and column of the matrix $\mathcal{A}_{p,\rho}$ by h/f and f/h, respectively, then we subtract the second row from the first, and finally to the first column we add the second column, multiplied by $b^{-1}a$. In the end we obtain a block-triangular matrix with the same determinant as $\mathcal{A}_{p,\rho}$, which proves (10.10). To complete the proof we apply Theorem 10.1.

$\square$

Let H_p, $1 < p < \infty$, denote the Hardy Banach space of all functions which are analytic in $\mathbb{D} = \{\varsigma \in \mathbb{C} : |\varsigma| < 1\}$, with the norm [28]

$$\|\varphi\| = \lim_{r \uparrow 1} \left(\int_0^{2\pi} |\varphi(re^{i\theta})|^p d\theta \right)^{1/p} \ .$$

We let h_p denote the Banach space, isometric to H_p, of numerical sequences $\xi = \{\xi_n\}_0^\infty$ which consists of Fourier coefficients of functions from H_p. With each function $a \in PC(\Gamma_0)$ we associate the operator T_a defined on h_p by the Toeplitz matrix $[a_{j-k}]_{j,k=0}^\infty$, where a_k are the Fourier coefficients of a. It is readily checked that T_a is the matrix representation of the operator $PaI \in \mathcal{L}(H_p)$, and hence $T_a \in \mathcal{L}(h_p)$.

Let $a = [a_{\ell m}] \in PC^{n \times n}(\Gamma_0)$. We set $T_a = [T_{a_{\ell m}}]_{\ell,m=1}^n$.

THEOREM 10.4. *Let $a \in PC^{n \times n}(\Gamma_0)$. Then T_a is a Φ_+- or Φ_--operator in h_p^n if and only if the matrix-valued function a is p-nonsingular. If this condition is satisfied, then*

$$Ind\ T_a = -ind\ W_p(a) \ .$$

PROOF. It follows from the definition of T_a that it is isometrically equivalent to the operator $Pa \in \mathcal{L}(h_p^n)$. Consequently, the operators $T_a(\in \mathcal{L}(h_p^n))$ and $B = PaP + Q(\in \mathcal{L}(L_p^n(\Gamma_0))$ are simultaneously normally solvable or not, $dim\ Ker\ T_a = dim\ Ker\ B$, and $dim\ Coker\ T_a = dim\ Coker\ B$. On the other hand, $B = (aP + Q)(I - QAP)$, and $I - QAP$ is invertible: $(I - QAP)^{-1} = I + QAP$. It remains to apply Theorem 10.1 to $aP + Q$.

□

Suppose that the contour Γ consists of a finite number of disjoint simple closed or open Lyapunov curves. Let τ_k, $k = 1, \ldots, m$ and τ_k, $k = m + 1, \ldots, 2m$ be the origin and the end points, respectively, of all open arcs of Γ. We let $\Lambda = L(\Gamma)$ denote the closed curve in $\mathbb{R}^3$ consisting of all the points (x, y, z) which satisfy the relations

$$x + iy \in \Gamma, \quad -1 \leq z \leq 1, \quad \text{and} \quad (1 - z^2) \prod_{k=1}^{2m} (x + iy - \tau_k) = 0 .$$

In other words, $\Lambda(\Gamma)$ is made of two copies of the curve Γ, lying in the planes $z = 1$ and $z = -1$, and also of all rectilinear segments which are parallel to the z-axis and pass through the extremities of the open arcs of Γ. In particular, if $\Gamma = [a, b]$, then $\Lambda(\Gamma)$ is the boundary of the rectangle $[a, b] \times [-1, 1]$.

We endow the curve $\Lambda(\Gamma)$ with the orientation which induces the original orientation of Γ (respectively, the one opposite the original orientation of Γ) on the copy of Γ in the plane $z = 1$ (respectively, $z = -1$). For each space $L_n(\Gamma, \rho)$, where

$$1 < p < \infty, \quad \rho(t) = \prod_{k=1}^{n} |t - t_k|^{\beta_k}, \quad -1 < \beta_k < p - 1 ,$$
$$n \geq 2m, \quad \text{and} \quad t_k = \tau_k , \quad k = 1, \ldots, 2m ,$$

we let $\Omega_{p,\rho}$ denote the function defined on $\Lambda(\Gamma)$ by the formulas

$$\Omega_{p,\rho}(t, z) = \begin{cases} \dfrac{z(t + d_k^2) - i(1 - z^2)d_k}{1 + z^2 d_k^2} & \text{for} \quad t = \tau_k, \ k \leq m \\[4mm] z & \text{for} \quad t \in \Gamma \setminus \{\tau_1, \ldots, \tau_{2m}\} \\[4mm] \dfrac{z(1 + d_k^2) + i(1 - z^2)d_k}{1 + z^2 d_k^2} & \text{for} \quad t = \tau_k , \ k \geq m + 1 , \end{cases} \tag{10.11}$$

where $d_k = cot(\pi(1 + \beta_k)/p)$.

THEOREM 10.5. *Suppose that Γ consists of a finite number of disjoint simple closed or open Lyapunov curves. Let $a, b \in C^{n \times n}(\Gamma)$. In order for $A = aI + bS$ to be a Φ_+- or Φ_--operator in $L^n_p(\Gamma, \rho)$ it is necessary and sufficient that*

$$det(a(t) + \Omega_{p,\rho}(t, z)b(t)) \neq 0 \qquad\qquad (10.12)$$

for all $(t, z) \in \Lambda(\Gamma)$.

If this condition is satisfied, then $A \in \Phi(E)$ and

$$Ind\ A = -ind\ det(a(t) + \Omega_{p,\rho}(t, z)b(t))\ . \qquad\qquad (10.13)$$

The proof of this theorem in the case $n = 1$ is given in [50(9), Sec.5, Chap. IX]. The proof in the general case is analogous.

$\square$

11. SUFFICIENT CONDITIONS FOR SINGULAR INTEGRAL OPERATORS WITH BOUNDED MEASURABLE COEFFICIENTS TO BE FREDHOLM OPERATORS

1. Let H be a separable Hilbert space. We let $L_\infty\left(\mathcal{L}(H)\right)$ denote the Banach space of weakly-measurable essentially bounded operator-valued functions $A : \Gamma_0 \to L(H)$, endowed with the norm

$$\|A\| = \operatorname*{ess\,sup}_{t \in \Gamma_0} \|A(t)\| \; .$$

Also, we let $L_\infty^+(\mathcal{L}(H))$ denote the subspace of $L_\infty(\mathcal{L}(H))$ consisting of the functions $A : \Gamma_0 \to L(H)$ which admit a bounded analytic extension $\tilde{A}$ to the interior of the unit disc:

$$\tilde{A}(t) = A(t) \; , \quad t \in \Gamma_0 \; ,$$

and

$$\sup_{|t|<1} \|\tilde{A}(t)\|_{L(H)} < \infty \; .$$

Let $\mathbf{A}$ be the operator of multiplication by the operator-valued function $A \in L_\infty(\mathcal{L}(H))$ acting in the space $L_p(\Gamma_0, H, \rho)$ (this space is defined in [73,Sec.7]):

$$(\mathbf{A}f)(t) = A(t)f(t) \; , \quad f \in L_p(\Gamma_0, H, \rho) \; .$$

We recall that the operator S of singular integration is bounded in the space $L_p(\Gamma_0, H, \rho)$ with the weight

$$\rho(t) = \prod_{k=1}^{n} |t - t_k|^{\alpha_k} \quad (t_k \in \Gamma_0 \; , \; 1 < p < \infty \; , \; -1 < \alpha_k < p-1)$$

(and so are the operators P and Q) (see Sec.7).

We next list a number of simple properties of the operators $\mathbf{A}P + \mathbf{B}Q$.

1. If $A \in L_\infty(L(H))$, then

$$\|\mathbf{A}\|_{L_p(\Gamma_0, H, \rho)} = \|A\|_{L_\infty} \; .$$

2. Let $A_1, B_1 \in L_\infty(\mathcal{L}(H))$ and $A_2, B_2^* \in L_\infty^+(\mathcal{L}(H))$. Then

$$(\mathbf{A}_1 P + \mathbf{B}_1 Q)(\mathbf{A}_2 P + \mathbf{B}_2 Q) = \mathbf{A}_1 \mathbf{A}_2 P + \mathbf{B}_1 \mathbf{B}_2 Q \ .$$

3. Suppose the operator-valued functions $A, B \in L_\infty(\mathcal{L}(H))$ take invertible values almost everywhere on Γ_0. If $A^{\pm 1}$, $(B^*)^{\pm 1} \in L_\infty^+(\mathcal{L}(H))$, then the operator $\mathbf{A}P + \mathbf{B}Q$ is invertible in $L_p(\Gamma_0, H, \rho)$, and

$$(\mathbf{A}P + \mathbf{B}Q)^{-1} = \mathbf{A}^{-1} P + \mathbf{B}^{-1} Q \ .$$

4. Suppose the operator-valued function $A \in L_\infty(\mathcal{L}(H))$ is of the form $A(t) = a(t)I$, where $a \in L_\infty(\Gamma_0)$. Then the operator $\mathbf{A}P + Q$ is invertible in $L_p(\Gamma_0, H, \rho)$ if and only if the operator $aP + Q$ is invertible in $L_p(\Gamma_0, \rho)$.

5. If the operator-valued function A belongs to $L_\infty(\mathcal{L}(H))$ and there is a $\delta > 0$ such that $Re \ A(t) \geq \delta$ for almost all $t \in \Gamma_0$, then the spectrum of the operator $\mathbf{A}$ is contained in the right half plane.

Properties 1 to 3 can be proved exactly as in the scalar case $H = \mathbb{C}$, 4 is verified directly, and for 5 the reader is referred to [25].

THEOREM 11.1. *Let $\rho(t) = |t - t_0|^\alpha$ with $t_0 \in \Gamma_0$, $-1 < \alpha < p - 1$, and $1 < p < \infty$. Let $A(t) = B_1^*(t) C(t) B_2(t)$, where $B_j^{\pm 1} \in L_\infty^+(\mathcal{L}(H))$, $j = 1, 2$, and $C \in L_\infty(\mathcal{L}(H))$. If*

$$\|I - C\|_{L_\infty} < sin\frac{\pi}{r} \ , \tag{11.1}$$

where

$$r = \max\{p, p(p-1)^{-1}, p(1+\alpha)^{-1}, p(p-1-\alpha)^{-1}\} \ ,$$

then the operator $\mathbf{A}P + Q$ is invertible in $L_p(\Gamma_0, H, \rho)$.

PROOF. It follows from property 2 that

$$\mathbf{A}P + Q = \mathbf{B}_1^*(CP + Q)[B_2 P + (\mathbf{B}_1^*)^{-1} Q] \ .$$

Since the factors $\mathbf{B}_1^*$ and $\mathbf{B}_2 P + (\mathbf{B}_1^*)^{-1} Q$ are invertible by property 3, it suffices to establish the invertibility of the operator $CP + Q$. To this end we need the following lemma.

LEMMA 11.1. *Let $R \in \mathcal{L}(H)$ and $2 < s < \infty$. Then the inequalities*

$$\|I - R\| \leq \sin\frac{\pi}{s} \qquad (11.2)$$

and

$$\|(I + \omega R)^{-1}(I - \omega R)\| \leq \tan\frac{\pi}{2s} , \qquad (11.3)$$

where

$$\omega = \left(\cos\frac{\pi}{s}\right)^{-1} ,$$

are equivalent.

PROOF OF THE LEMMA. Condition (11.2) is equivalent to

$$(I - R)(I - R^*) \leq \sin\frac{2\pi}{s} I$$

or, equivalently, to

$$\omega^{-2} I + RR^* \leq R + R^* , \qquad (11.4)$$

which in turn is the same as

$$(I - \omega R)(I - \omega R^*) \leq \left(\tan\frac{\pi}{2s}\right)^2 (I + \omega R)(I + \omega R^*) . \qquad (11.5)$$

Inequality (11.2) guarantees the invertibility of the operator $I + \omega R$. In fact, $I + \omega R = (1 + \omega)[I + \omega(1 + \omega)^{-1}(R - I)]$ and $\|\omega(1 + \omega)^{-1}(R - I)\| \leq \tan\frac{\pi}{2s} < 1$. Multiplying both sides of inequality (11.4) by $(I + \omega R)^{-1}$ on the left and by $(I + \omega R^*)^{-1}$ on the right, we get

$$(I + \omega R)^{-1}(I - \omega R)(I - \omega R^*)(I + \omega R^*)^{-1} \leq \left(\tan\frac{\pi}{2s}\right)^2 I , \qquad (11.6)$$

which is equivalent to (11.3).

$\square$

We now return to the proof of Theorem 11.1. It follows from (11.1) and (11.4) that for some $s > r$

$$C(t) + C^*(t) \geq (cos\frac{\pi}{s})^2 I \ .$$

Set $\omega = (cos \ \pi/s)^{-1}$. By property 5, the operator $I + \omega C$ is invertible in $L_p(\Gamma_0, H, \rho)$. We write

$$CP + Q = \frac{1}{2}(I + \omega C)[I + (I + \omega C)^{-1}(I - \omega C)S_0](\omega^{-1}P + Q) \ . \tag{11.7}$$

Since $(\omega^{-1}P + Q)(\omega P + Q) = I$, it remains to verify that the operator $X = I + (I + \omega C)^{-1}(I - \omega C)S_0$ is invertible. By Theorem 8.1, $\|S_0\| = cot\frac{\pi}{2r}$. Moreover, by Lemma 11.1 there is an $S > r$ such that

$$\|(I + \omega C)^{-1}(I - \omega C)\|_{L_p(\Gamma_0, H, \rho)} \leq tan\frac{\pi}{2s} \ .$$

Consequently, $\|I - X\| < 1$, which proves the invertibility of X.

□

THEOREM 11.2. *The assertion of Theorem 11.1 remains valid when condition (11.1) is replaced by the following ones:*

$$\left. \begin{array}{c} \underset{t \in \Gamma_0}{ess \ inf} \ \|C(t)\| > 0 \\ and \\ Re\frac{C(t)}{\|C(t)\|} > \delta I \ > cos\frac{\pi}{r}I \ . \end{array} \right\} \tag{11.8}$$

PROOF. Set $D(t) = cos\frac{\pi}{r}\|C(t)\|^{-1}C(t)$. Then obviously $Re \ D(t) \geq \delta \ cos\frac{\pi}{r}I > (cos\frac{\pi}{r})^2 I$. We claim that the operator-valued function D satisfies condition (11.1) or, equivalently, the inequality

$$(cos\frac{\pi}{s})^2 I + D(t)D^*(t) \leq D(t) + D^*(t) \tag{11.9}$$

for some $s > r$. In fact, for s sufficiently close to r

$$\begin{aligned} (cos\frac{\pi}{s})^2 I + D(t)D^*(t) &= (cos\frac{\pi}{s})^2 I + (cos\frac{\pi}{r})^2\|C(t)\|^{-2}C(t)C^*(t) \\ &\leq [(cos\frac{\pi}{s})^2 + (cos\frac{\pi}{r})^2]I \ < \ 2\delta \ cos\frac{\pi}{r}I \ \leq \ D(t) + D^*(t) \ . \end{aligned}$$

Since ess inf$\|C(t)\| > 0$ for $t \in \Gamma_0$, the function $a(t) = \|C(t)\|$ admits a factorization $a = a_+\bar{a}_+$, where $a_+^{\pm 1} \in L_\infty(\Gamma_0)$ (see [50(9)]). Therefore, the operator-valued function $C(t) = (cos\frac{\pi}{r})^{-1}a(t)D(t)$ satisfies the conditions of Theorem 11.1.

□

<u>Remark 11.1.</u> The constants $sin\frac{\pi}{r}$ and $cos\frac{\pi}{r}$ in the formulations of Theorems 11.1 and 11.2 are exact: for smaller values of the constants the indicated results are not valid. In fact, consider the operator-valued function $C(t) = exp(i\beta(\theta - \theta_0 - \pi))I$, where $t = exp(i\theta)$, $\tau = exp(i\theta_0)$, and $\theta_0 \le \theta < \theta_0 + 2\pi$. Let $\rho(t) = |t - t_0|^\alpha$. By property 4, the operator $CP + Q$ is invertible in $L_p(\Gamma_0, H, \rho)$ if and only if $exp(i\beta(\theta - \theta_0 - \pi))P + Q$ is invertible in $L_p(\Gamma_0, \rho)$. We set

$$\tau = t_0, \quad \beta = 1/r \quad \text{for} \quad r = \frac{p}{1 + \alpha} \quad ,$$

$$\tau = t_0, \quad \beta = -1/r \quad \text{for} \quad r = \frac{p}{p - 1 - \alpha} \quad ,$$

$$\tau \ne t_0, \quad \beta = 1/r \quad \text{for} \quad r = p \quad ,$$

$$\tau \ne t_0, \quad \beta = -1/r \quad \text{for} \quad r = \frac{p}{p - 1} \quad .$$

By Theorem 9.1, the operator $exp(i\beta(\theta - \theta_0 - \pi))P + Q$ is not invertible in $L_p(\Gamma_0, \rho)$. But

$$Re\ C(t)\|C(t)\|^{-1} = cos\frac{\theta - \theta_0 - \pi}{r}I \ge cos\frac{\pi}{r}I \ .$$

2. Let Γ be a simple closed Lyapunov contour, $t_1, \ldots, t_n$ distinct points on Γ, and $\rho(t) = \Pi|t - t_k|^{\alpha_k}$, where $-1 < \alpha_k < p - 1$, $1 < p < \infty$.

We let $S^n_{p,\rho}(\Gamma)$ denote the class of matrix-valued functions $A \in L^{n \times n}_\infty(\Gamma)$ which satisfy the following condition:

- *for every point $\tau \in \Gamma$ there exists a neighborhood $U(\tau)$ and a pair of matrix-valued functions $B_1, B_2 \in GL^+_\infty(\Gamma)^{n \times n}$ such that the matrix-valued function $C(t) = B_1^*(t)A(t)B_2(t)$ is subject to the bound*

$$\underset{t \in U(\tau)}{\text{ess sup}} \|I - C(t)\|_{\mathbb{C}^{n \times n}} < sin\frac{\pi}{r(\tau)} \ ,$$

where

$$r(\tau) = \begin{cases} \max(p, p(p-1)^{-1}) & \text{if} \quad \tau = t_k, \ k = 1, \ldots, m \\ \max(p, p(p-1)^{-1}, p(1+\alpha_k)^{-1}, p(p-1-\alpha_k)^{-1}) & \text{if} \quad \tau = t_k \ . \end{cases}$$

THEOREM 11.3. *If $A \in S_{p,\rho}^n(\Gamma)$, then $AP + Q$ is a Φ-operator in $L_p^n(\Gamma, \rho)$.*

This theorem is proved with the help of a local principle, exactly as in the case $n = 1$ (see, e.g., [50(9)]). In the proof one uses the norm of the operator S, calculated in Theorem 11.1.

□

Remark 11.2. In general, Theorem 11.3 fails when the space H is infinite dimensional. To see this, choose two functions $a, b \in S_{p,\rho}^1$ (they may even be continuous) such that if $A = aP + Q$ and $B = bP + Q$ then $dim \ Ker \ A > 0$ and $dim \ Coker \ B > 0$. Consider the diagonal operator $\tilde{A} = diag(A, B, A, B, \ldots)$ acting in $L_p(\Gamma, H, \rho)$. It is readily checked that $dim \ Ker \tilde{A} = \infty$ and $dim \ Coker \ \tilde{A} = \infty$.

Remark 11.3. Let $\rho(t) \equiv 1$ and $p \geq 2$. It follows from the definition of the class S_p^n $(=S_{p,1}^n)$ that $S_r^n \subset S_p^n$ for every $r \in [q, p]$, where $q^{-1} + p^{-1} = 1$. Therefore, if $a \in S_p^n(\Gamma)$, then the operator $A = aP + Q \in \Phi(L_r^n(\Gamma))$. It follows from the results of [69] that in this case $Ind \ A$ is the same in all spaces $L_r^n(\Gamma)$ with $r \in [q, p]$.

Remark 11.4. The condition $a \in S_{p,\rho}^n$ is not necessary for the operator A to belong to $\Phi(L_p^n(\Gamma, \rho))$. Take, for example, $n = 1$, $p = 4$, and $a(t) = t^{1/2}$. It follows from Theorem 9.1 that $aP + Q \in \Phi(L_4(\Gamma))$, but $aP + Q \notin \Phi(L_2(\Gamma))$. By Remark 11.3, $a \notin S_4^1(\Gamma)$.

It turns out that the condition $a \in S_p^1(\Gamma)$ is necessary and sufficient in order for the operator A to have the properties: $A \in \Phi(L_p(\Gamma)) \cap \Phi(L_q(\Gamma))$ and $Ind \ A|_{L_p} = Ind \ A|_{L_q}$. This assertion will be proved in the next section.

12. NECESSARY CONDITIONS FOR FREDHOLMNESS

Let Γ_0 be a simple closed Lyapunov contour. We let $S_p(\Gamma)$ (with $p \geq 2$) denote the set of functions $a \in GL_\infty(\Gamma)$ which can be represented in the form $a = hg$, where $h \in GL_\infty^+(\Gamma)$ and the range of the function g lies in a sector $Y(g)$ of opening less than $2\pi/\rho$ with vertex at the point $\lambda = 0$.

It is readily verified that $S_p(\Gamma) \subset S_p^1(\Gamma)$. It follows from Remark 11.3 that if $a \in S_p(\Gamma)$, then $A = aP + Q \in \Phi(L_r(\Gamma))$ for all $r \in [q, p]$, where $q^{-1} + p^{-1} = 1$. Let ω be the complex point at which the bisector of $Y(g)$ intersects the unit circle Γ_0. Then for every choice of $\lambda \in [0, 1]$, the function $a_\lambda = \lambda a + (1 - \lambda \omega) \in S_p(\Gamma)$, and hence $A_\lambda = a_\lambda P + Q \in \Phi(L_r(\Gamma))$. Consequently, $Ind\ A = Ind\ A_\lambda$ $(0 \leq \lambda \leq 1)$. But $(\omega P + Q)(\omega^{-1} P + Q) = I$, whence $Ind\ A = 0$. By Theorem 9.2, $A \in G\mathcal{L}(L_r(\Gamma))$ for all $r \in [q, p]$.

THEOREM 12.1. *Let $a \in L_\infty(\Gamma)$. The operator $A = aP + Q$ is invertible in the spaces $L_p(\Gamma)$ and $L_q(\Gamma)$ (where $p^{-1} + q^{-1} = 1$, $p \geq 2$) if and only if $a \in S_p(\Gamma)$.*

PROOF. The sufficiency of this condition has been established above. To prove its necessity we need an auxiliary result which generalizes the well-known Helson-Szegö criterion for the boundedness of the operator S in spaces $L_2(\Gamma, \rho)$ (see [26], and also [34]).

In this section (and only here) we deviate from the definition of the norm in $L_p(\Gamma, \rho)$ generally used in the book. Here we set

$$\|f\|_{L_p(\Gamma, \rho)} = \|\rho f\|_{L_p(\Gamma)} . \tag{12.1}$$

With this definition the Hunt-Muckenhoupt-Wheeden criterion [29, 11] for the boundedness of the operator S in $L_p(\Gamma_0, \rho)$ takes the symmetric form

$$\rho \in W_p(\Gamma_0) \iff \left(\int_\ell \rho^p(t) |dt| \right)^{1/p} \left(\int_\ell \rho^{-q}(t) |dt| \right)^{1/q} \leq A(p) |\ell| \tag{12.2}$$

for every arc $\ell \subset \Gamma_0$ (of length $|\ell|$; the constant $A(p)$ does not depend on ℓ); here $W_p(\Gamma)$ designates the set of all measurable nonnegative functions ρ on Γ which are different from zero a.e. and such that the operator of singular integration is bounded in $L_p(\Gamma_0, \rho)$.

THEOREM 12.2. *Let $p \geq 2$ and $q = p(p-1)^{-1}$. Then*

$$\rho \in W_p(\Gamma_0) \cap W_q(\Gamma_0) \iff (\rho(e^{i\theta}) = exp(n(\theta) + \tilde{v}(\theta))) , \qquad (12.3)$$

where u and v are real-valued functions from $L_\infty(0, 2\pi)$,

$$ess\,sup\,|v(\theta)| < \pi/2p , \qquad (12.4)$$

and

$$\tilde{v}(\theta) = \frac{1}{2\pi} \int_0^{2\pi} cot\frac{\sigma - \theta}{2} v(\sigma) d\sigma .$$

PROOF. The implication $\Leftarrow$ is well known ([70]; see also [12]). Let us prove the converse. Let $\rho \in W_p(\Gamma_0) \cap W_q(\Gamma_0)$ and let $\ell \subset \Gamma_0$ be an arbitrary arc. Then it follows from the criterion (12.2) that

$$\left(\int_\ell \rho^p(t)|dt| \right)^{1/p} \left(\int_\ell \rho^{-q}(t)|dt| \right)^{1/q} \leq A(p)|\ell| \qquad (12.5)$$

and

$$\left(\int_\ell \rho^q(t)|dt| \right)^{1/q} \left(\int_\ell \rho^{-p}(t)|dt| \right)^{1/p} \leq A(q)|\ell| . \qquad (12.6)$$

Next, from (12.5), (12.6), and the Cauchy-Bunyakovskii inequality

$$|\ell|^2 \leq \int_\ell \rho^q(t)|dt| \int_\ell \rho^{-q}(t)|dt|$$

it follows that

$$\left(\int_\ell \rho^p(t)|dt| \right)^{1/2} \left(\int_\ell \rho^{-p}(t)|dt| \right)^{1/2} \leq |\ell|(A(p)A(q))^{p/2} . \qquad (12.7)$$

By (12.2), $\rho^{p/2} \in W_2(\Gamma))$. Now we can apply the Helson-Szegö criterion for the boundedness of the operator S in $L_2(\Gamma, \rho)$ ([26]; see also [34]) and conclude that $\rho^{p/2}(e^{i\theta}) =$

$exp(u_1(\theta) + \tilde{v}_1(\theta))$, where u_1 and v_1 are bounded real-valued functions on $[0, 2\pi]$ and ess $\sup|v_1(\theta)| < \pi/4$. To complete the proof of Theorem 12.2 we take $u = 2u_1/p$ and $v = 2v_1/p$.

$\square$

Returning to the proof of Theorem 12.1, suppose that the operator $A = aP + Q$ is invertible in L_p and L_q. We show that $a \in S_p$. By Theorems 9.5—9.7, it suffices to examine the case where $\Gamma = \Gamma_0$ and $|a(t)| = 1$. The function a admits p- and q-factorizations $a = a_+ a_-$ and $a = b_+ b_-$, respectively (Theorem 9.4). From the equality $a_+ a_- = b_+ b_-$ it follows that $a_+ b_+^{-1} = b_- a_-^{-1} \in L_1^+(\Gamma_0) \cap L_1^-(\Gamma_0)$, and hence that $a_+ = \gamma b_+$ and $a_- = \gamma^{-1} b_-$, where $\gamma =$ const. Also, it follows from the definition of p- and q-factorizations that $a_+^{\pm 1} \in L_p^+(\Gamma_0)$, $a_-^{\pm 1} \in L_p^-(\Gamma)$, and $|a_+^{-1}| \in W_p(\Gamma_0) \cap W_q(\Gamma_0)$. Moreover, $a_+ \bar{a}_+ a_- \bar{a}_- = 1$, and hence $a_+ \bar{a}_- = a_+^{-1} \bar{a}_+^{-1} \in L_{p/2}^+(\Gamma_0) \cap L_{p/2}^-(\Gamma_0)$. Consequently, $a_- = \gamma_1/\bar{a}_+$, where $\gamma_1 \in \mathbb{C}$ and $|\gamma_1| = 1$. Multiplying, if necessary, a_- by γ_2 and a_+ by γ_2^{-1}, with $\gamma_2^{-2} = \gamma_1$, we may assume that $a_- = \bar{a}_+^{-1}$. The space $L_p^+(\Gamma_0)$ coincides with the Hardy space H_p [28]. The functions a_+ and a_+^{-1} admit factorizations $a_+ = g_0 b$ and $a_+^{-1} = h_0 d$, where g_0 and h_0 are inner functions, while b and d are outer:

$$b(z) = exp\left\{\frac{1}{2\pi}\int_{-\pi}^{\pi}\frac{e^{i\theta} + z}{e^{i\theta} - z}\ell n|a_+(e^{i\theta})|d\theta\right\}$$

and

$$d(z) = exp\left\{\frac{1}{2\pi}\int_{-\pi}^{\pi}\frac{e^{i\theta} + z}{e^{i\theta} - z}\ell n|a_+^{-1}(e^{i\theta})|d\theta\right\} = 1/b(z)$$

(see [28]). Therefore, $1 = g_0 h_0$. Since g_0 and h_0 are inner, they must be constants. We thus conclude that $a_+^{\pm 1}$ are outer functions, and hence

$$a_+ = \lambda \, exp(2P\ell n|a_+|) , \qquad (12.8)$$

where $P = \frac{1}{2}(I + S_0)$ and $\lambda \in \mathbb{C}$. Moreover, $a_+^{-1} \in W_p(\Gamma_0) \cap W_q(\Gamma_0)$ and, by Theorem 12.2, $|a_+|^{-1} = exp(u + \tilde{v})$, where $u, v \in L_\infty(\Gamma_0)$ are real-valued and $\|v\|_\infty < \pi/2p$. It is readily checked that if the function φ is real-valued, then $\overline{S_0\varphi} + S_0\varphi = $ const. Since

$\tilde{v} = iS_0 v + \text{const.}$, it follows that

$$a_+ = \mu \ exp(-2P(u - iS_0 v)) = \mu \ exp(-u - S_0 u + iv + iS_0 v)$$

and

$$a_- = \bar{a}_+^{-1} = \mu_1 exp(u - S_0 u + iv - iS_0 v) \ .$$

Therefore, $a = a_+ a_- = \nu \ exp(-2S_0 u + 2iv) = \nu \ exp(-4Pu + 2u + 2iv)$, with $\nu \in \mathbb{C}$. If we now set $h = \nu \ exp(-4Pu)$ and $g = exp(2u + 2iv)$, then $a = hg$. By Theorem 9.7, $h \in GL_\infty^+$, and hence h and g satisfy the conditions that define the class S_p.

$\square$

COROLLARY 12.1. *Let $a \in L_\infty(\Gamma)$. If the operator $A = aP + Q$ is Fredholm in both $L_p(\Gamma)$ and $L_q(\Gamma)$ and if $Ind \ A|_{L_p} = Ind \ A|_{L_q} = \kappa$, then the function a can be expressed as*

$$a(t) = (hg)(t)(t - z_0)^{-\kappa} \ ,$$

where h and g satisfy the conditions that define the class S_p and z_0 belongs to the domain bounded by the contour Γ.

PROOF. The operator A admits the representation

$$A = ((t - z_0)^\kappa aP + Q)((t - z_0)^{-\kappa}P + Q) + T \ ,$$

where $T \in \mathcal{T}(E)$. Since

$$Ind((t - z_0)^{-\kappa}P + Q) = \kappa \ ,$$

we have

$$Ind((t - z_0)^\kappa aP + Q) = 0 \ .$$

It follows from Theorem 9.2 that the operator $(t - z_0)^\kappa aP + Q$ is invertible. By Theorem 12.1, $(t - z_0)^\kappa a = hg$, i.e. $a = hg(t - z_0)^{-\kappa}$, as claimed.

$\square$

COROLLARY 12.2. *Let $a \in L_\infty$. Then $aP + Q$ is a Φ-operator in every space $L_r(\Gamma)$ with $q \leq r \leq p$ if and only if $a \in S_p(\Gamma)$.*

PROOF. If $A \in \Phi(L_r(\Gamma))$ for every $r \in [q, p]$, then $Ind\ A$ is the same in all these spaces [69]. Hence, by Corollary 12.1., $a = hg(t - z_0)^{-\kappa}$. Since the values of the function g lie in a sector of opening less than $2\pi/p$, for every point $\tau \in \Gamma$ there is a neighborhood $U(\tau)$ such that for $t \in U(\tau)$ the values of the function $g(t)(t - z_0)^{-\kappa}$ also lie in a sector of opening less than $2\pi/p$.

$\square$

CHAPTER IV

BANACH ALGEBRAS WITH SYMBOL

In this chapter we characterize the Banach algebras possessing a scalar symbol and give examples of such algebras.

13. SUFFICIENT FAMILIES OF MULTIPLICATIVE FUNCTIONALS

Let K be a Banach algebra with identity e and let GK denote the group of invertible elements of K. We recall that the radical $\mathcal{R}(K)$ of algebra K is by definition the intersection of all its left maximal ideals. It is known that the element x_0 belongs to $\mathcal{R}(K)$ if and only if $e + ux_0 \in GK$ (or $e + x_0u \in GK$) for all $u \in K$ (see [62]). The radical $\mathcal{R}(K)$ is a closed two-sided ideal of K. We set $\tilde{K} = K/\mathcal{R}(K)$ and we denote by $\tilde{x}$ the element of the quotient algebra $\tilde{K}$ which contains x. We shall need the following well-known properties of Banach algebras.

13.1. Let K be a commutative Banach algebra, $\mathcal{M}$ the set of its two-sided maximal ideals, and f_M the multiplicative functional on K corresponding to the ideal $M \in \mathcal{M}$ (i.e., $Ker\, f_M = M$). Then

$$x \in GK \Longleftrightarrow f_M(x) \neq 0 \ \text{ for all } \ M \in \mathcal{M}$$

(see [20]).

13.2. If the complement of the spectrum of the element $x \in K$ is connected, then the spectrum of x does not change on passing to any subalgebra of K (see [7]).

LEMMA 13.1. *If $x \in K$, then $x \in GK$ if and only if $\tilde{x} \in G\tilde{K}$.*

PROOF. Let $\tilde{x} \in G\tilde{K}$. Then there is a $z \in K$ such that $xz = e + u_1$ and $zx = e + u_2$, where $u_1, u_2 \in \mathcal{R}(K)$. Since $e + u_k \in GK$, it follows that $x \in GK$. The converse is obvious.

$\square$

Definition 13.1. A family $\{f_M\}_{M \in \alpha}$ of multiplicative functionals on the Banach algebra K is said to be *sufficient* if it has the following property:

$$x \in GK \iff f_M(x) \neq 0 \text{ for all } M \in \alpha. \tag{13.1}$$

If the Banach algebra K possesses a sufficient family of multiplicative functionals, then we shall write $K \in \pi_1$.

Every commutative Banach algebra K belongs to the class π_1 since, by property 13.1, the set of all multiplicative functionals $\{f_M\}_{M \in \mathcal{M}}$ is a sufficient family.

A simple example of noncommutative Banach algebra $K \in \pi_1$ is provided by the algebra of upper-triangular 2×2 matrices with scalar entries:

$$K = \left\{ a = \begin{pmatrix} a_{11} & a_{12} \\ 0 & a_{22} \end{pmatrix} : a_{jk} \in \mathbb{C} \right\}.$$

In fact, the two functionals: $f_1(a) = a_{11}$ and $f_2(a) = a_{22}$ form a sufficient family. In this example

$$\mathcal{R}(K) = \left\{ \begin{pmatrix} 0 & a_{12} \\ 0 & 0 \end{pmatrix} : a_{12} \in \mathbb{C} \right\}$$

and the quotient algebra $\tilde{K}$ is commutative. It turns out that this last property is characteristic for all algebras $K \in \pi_1$.

THEOREM 13.1. *The algebra K possesses a sufficient family of multiplicative functionals if and only if the quotient algebra $\tilde{K} = K/\mathcal{R}(K)$ is commutative.*

PROOF. Suppose that $\tilde{K}$ is commutative and let $\{\tilde{f}_M\}$ be the set of all multiplicative functionals on $\tilde{K}$. We set $f_M(x) = \tilde{f}_M(\tilde{x})$. Then, by Lemma 13.1 and property 13.2,

$$x \in GK \iff \tilde{x} \in G\tilde{K} \iff (\tilde{f}_M(\tilde{x}) \neq 0 \text{ for all } M) \iff (f_M(x) \neq 0 \text{ for all } M),$$

i.e., $\{f_M\}$ is a sufficient family for K.

Conversely, suppose $K \in \pi_1$ and let $\{f_M\}_{M \in \alpha}$ be a sufficient family of multiplicative functionals on K. Pick an arbitrary pair of elements $x, y \in K$, and let us show that $z = [x, y] = xy - yx$ belongs to $R(K)$. For every $u \in K$ and every $M \in \alpha$,

$$f_M(e + uz) = 1 + f_M(u)(f_M(x)f_M(y) - f_M(y)f_M(x)) = 1 \, ,$$

whence $e + uz \in GK$, i.e., $z \in R(K)$. We have thus shown that $\tilde{K}$ is commutative.

$\square$

COROLLARY 13.1.   *A semisimple Banach algebra K belongs to π_1 if and only if it is commutative.*

$\square$

COROLLARY 13.2.   *A Banach algebra K belongs to π_1 if and only if the quotient algebra $K/R(K)$ is isometrically isomorphic to a Banach algebra $\mathring{K}$ of continuous functions $\mathring{x}: M\rangle \mathbb{C}$ on a (Hausdorff) compact topological space M.*

Definition 13.2.   An element $z \in K$ is called an *admissible perturbation for invertible elements* if $a + z \in GK$ for every element $a \in GK$.

THEOREM 13.2.   *The algebra K possesses a sufficient family of multiplicative functionals if and only if every commutator $z = [a, b], \quad a, b \in K$, is an admissible perturbation for invertible elements.*

PROOF.   Suppose that $K \in \pi_1$ and let $c \in GK$, $a, b \in K$, and $z = [a, b]$. Then $z \in R(K)$, and hence $c + z \in GK$. To prove the converse it suffices to show that every admissible perturbation $z \in K$ belongs to the radical of K. Let $u \in K$ and let z be an admissible perturbation. Then

$$e + uz = (u - \lambda e)[(u - \lambda e)^{-1} + z] + \lambda z \, ,$$

where λ is a regular point for u. Consequently, $e + uz \in GK$, i.e., $z \in R(K)$.

□

We next consider several examples.

Example 13.1. Let E be a Banach space. If $dim\ E > 1$, then $\mathcal{L}(E) \notin \pi_1$. In fact, if $dim\ E > 1$ the algebra $\mathcal{L}(E)$ is not commutative, and our assertion follows from Corollary 13.1 and the following lemma:

LEMMA 13.2. *Algebra $\mathcal{L}(E)$ is semisimple for every Banach space E.*

PROOF. Let $A \in \mathcal{L}(E)$, $A \neq 0$. We show that $A \notin R(\mathcal{L}(E))$. Since $A \neq 0$, there exists an $x_0 \in E$ such that $Ax_0 = y_0 \neq 0$. Let f be a continuous functional on E such that $f(y_0) = 1$, and consider the operator $B \in \mathcal{L}(E)$ defined by the rule $Bx = f(x)x_0$. Then $(I - BA)x_0 = 0$, and hence $I - BA \notin G\mathcal{L}(E)$. Therefore, $A \notin R(L(E))$, as claimed.

□

Example 13.2. Let $A \in \mathcal{L}(E)$ and $A^{m+1} = 0$ for some $m \in \mathbb{N}$. Set

$$K = \{\alpha_0 I + \alpha_1 A + \ldots + \alpha_m A^m : \alpha_j \in \mathbb{C}\} \subset \mathcal{L}(E) .$$

In this example $R(K) = \{\alpha_1 A + \ldots + \alpha_m A^m : \alpha_j \in \mathbb{C}\}$, the algebra $\tilde{K}$ is commutative, and hence $K \in \pi_1$. A sufficient family of multiplicative functionals is provided by the single functional f : $f(\alpha_0 I + \alpha_1 A + \ldots + \alpha_m A_m) = \alpha_0$. The operator $B = \alpha_0 I + \ldots + \alpha_m A^m$ is invertible in K if and only if $\alpha_0 \neq 0$. We remark that the condition $\alpha_0 \neq 0$ is the criterion for the invertibility of B in $\mathcal{L}(E)$.

Example 13.3. Let $\Gamma_0 = \{z \in \mathbb{C} : |z| = 1\}$, the unit circle, $E = L_p(\Gamma_0)$, $C_+(\Gamma_0)$ the set of all functions analytic in the open unit disc $\mathbb{D} = \{z \in \mathbb{C} : |z| < 1\}$ and continuous on $\overline{\mathbb{D}}$, and K the subalgebra of $\mathcal{L}(E)$ consisting of all operators H of multiplication by functions $h \in C_+(\Gamma_0)$: $(Hx)(t) = h(t)x(t)$. In this example every multiplicative functional on K is specified by a point $\tau \in \overline{\mathbb{D}}$: $f_\tau(H) = h(\tau)$.

The conditions for the invertibility of an element H in the algebras K and $\mathcal{L}(E)$ are not identical. For example, the operator H_0 of multiplication by the independent variable is invertible in $\mathcal{L}(E)$, but not in K. We recall that a subalgebra K is said to be *inverse-closed* (or *to contain all inverses*) in the algebra K° if $x^{-1} \in K$ for every element $x \in GK^\circ \cap K$.

In Example 13.3 K is an inverse-closed subalgebra in $\mathcal{L}(E)$.

<u>Definition 13.3.</u> The sufficient family $\{f_M\}_{M \in \alpha}$ of multiplicative functionals is called *symmetric* if for every $x \in K$ there is an $\bar{x} \in K$ such that $f_M(\bar{x}) = \overline{f_M(x)}$ for all $M \in \alpha$.

THEOREM 13.3.   *Let K° be a Banach algebra and let K be a commutative subalgebra of K° which possesses a symmetric sufficient family of multiplicative functionals. Then K is inverse-closed in K°.*

To prove this we need the following lemma:

LEMMA 13.3.   *Let $x, y \in K^\circ$ be such that $xy = yx$ and the spectra of x and y in K° are real. If $x + iy \in GK^\circ$, then $x - iy \in GK^\circ$.*

PROOF.   Suppose that $xy = yx$ and $x + iy \in GK^\circ$. We let K_1 denote the smallest Banach subalgebra with identity of K° which contains the elements x, y, and $(x+iy)^{-1}$. Since $(x+iy)^{-1}x = x(x+iy)^{-1}$ and $(x+iy)^{-1}y = y(x+iy)^{-1}$, the subalgebra K_1 is commutative. Let $\mathcal{M}$ denote the set of maximal ideals of K_1. Since $x + iy \in GK_1$, $(x+iy)(M) \neq 0$ for every $M \in \mathcal{M}$ (where we put $z(M) = f_M(z)$ for $z \in K_0$ and $M \in \mathcal{M}$). The spectra of x and y in K° are real by assumption. By property 13.2, the spectra of these elements in K_1 are also real. But the ranges of the functions $x(M)$ and $y(M)$ on $\mathcal{M}$ coincide with the spectra of x and y, respectively, in K_1. Consequently, $x(M) \in \mathbb{R}$ and $y(M) \in \mathbb{R}$ for all $M \in \mathcal{M}$, and hence $(x - iy)(M) = \overline{(x+iy)(M)} \neq 0$ for all $M \in \mathcal{M}$. Therefore, $x - iy \in GK_1 \subset GK^\circ$, as claimed.

□

PROOF OF THEOREM 13.3.   Let $x \in K \cap GK^\circ$. Since K possesses a symmetric sufficient family $\{f_M\}_{M \in \alpha}$ of multiplicative functionals, there is an element $\bar{x} \in K$ such that $f_M(\bar{x}) = \overline{f_M(x)}$ for all $M \in \alpha$. Let $a = (x + \bar{x})/2$ and $b = (x - \bar{x})/2i$.

Then $f_M(a - \lambda e) \neq 0$ and $f_M(b - \lambda e) \neq 0$ for every $\lambda \in \mathbb{C} \setminus \mathbb{R}$, and hence the spectra of the elements a and b are real. Moreover, $ab = ba$ and $a + ib = x \in GK^\circ$. By Lemma 4.3, $\bar{x} = a - ib \in \mathcal{G}K^\circ$. Consequently, $x\bar{x} \in GK^\circ$. But the spectrum of $x\bar{x}$ in K is real: in fact, if $\lambda \in \mathbb{C} \setminus \mathbb{R}$, then $f_M(x\bar{x} - \lambda e) \neq 0$ for every $M \in \alpha$. Hence, it coincides with the spectrum of $x\bar{x}$ in K°, which implies that $x\bar{x} \in GK$. This in turn implies, as is readily checked, that $x \in GK$.   □

Example 13.4.   Let $A \in \mathcal{L}(E)$, $A^{m+1} = I$ for an $m \in \mathbb{N}$, and let K be the algebra generated by the operator A:

$$K = \{\alpha_0 I + \alpha_1 A + \ldots + \alpha_m A^m : \alpha_j \in \mathbb{C}\}.$$

Then K is commutative, $f(A^k)f(A^{m+1-k}) = 1$ and $|f(A^k)| = 1$ for every multiplicative functional f. Consequently, $f(A^{m+1-k}) = \overline{f(A^k)}$ and, more generally, $f(\Sigma\bar{\alpha}_k A^{m+1-k}) = \overline{f(\Sigma\alpha_k A^k)}$, i.e., algebra K is symmetric. The multiplicative functionals

$$f_p\left(\sum_{k=0}^{m} \alpha_k A^k\right) = \sum_{k=0}^{m} \alpha_k \varepsilon_p^k,$$

where $\varepsilon_p \in \sigma(A) = $ the spectrum of A, form a sufficient family. If $B = \sum_{k=0}^{m} \alpha_k A^k$, then

$$B \in GK \Longleftrightarrow B \in G\mathcal{L}(E) \Longleftrightarrow f_p(B) \neq 0 \quad \forall p.$$

Example 13.5.   Let $E = L_2(a,b)$, and let K denote the subalgebra of $\mathcal{L}(E)$ generated by the singular integral operator S:

$$(S\varphi)(t) = \frac{1}{\pi i} \int_a^b \frac{\varphi(\tau)d\tau}{\tau - t}, \quad t \in [a,b]. \tag{13.2}$$

Since $S^* = S$, K is a C^*-subalgebra of $\mathcal{L}(E)$; in particular, K is symmetric. The spectrum of the element S in K coincides with its spectrum in $\mathcal{L}(E)$, i.e., with the segment $[-1,1]$ (see [50(9)]). Every multiplicative functional is defined by a point $\mu \in [-1,1]$:

$$f_\mu(\Sigma_{k=0}^{N}\alpha_k S^k) = \Sigma_{k=0}^{N}\alpha_k\mu^k. \tag{13.3}$$

In particular, if $A = \alpha I + \beta S$ (where $\alpha, \beta \in \mathbb{C}$), then

$$A \in GK \iff A \in G\mathcal{L}(E) \iff (\alpha + \beta\mu \neq 0 \quad \forall\, \mu \in [-1,1])\ .$$

Consider the operator B defined by the formula

$$(B\varphi)(t) = \alpha\varphi(t) + \frac{\beta}{\pi i} \int_a^b \frac{\varphi(\tau)d\tau}{\tau - t} + \frac{\gamma}{(\pi i)^2} \int_a^b \ell n \frac{(b-t)(\tau-a)}{(t-a)(b-\tau)}\, \frac{\varphi(\tau)d\tau}{\tau - t}\ . \tag{13.4}$$

Then $B \in K$. In fact it is readily checked, using the Poincaré-Bertrand formula [18], that $B = \alpha I + \beta S + \gamma(S^2 - I)$. We are thus led to the following result:

THEOREM 13.4. *The operator B in $L_2(a,b)$ defined by formula (13.4) is invertible if and only if $\gamma\tau^2 + \beta\tau + \alpha - \gamma \neq 0$ for every $\tau \in [0,1]$.*

$\square$

Example 13.6. Let $E = L_p(\mathbb{R}^+, t^\beta)$, where $\mathbb{R}^+ = [0,\infty)$, $1 < p < \infty$, $-1 < \beta < p-1$, with the norm $\|x\|_E = \|xt^{\beta/p}\|_{L_n(\mathbb{R}+)}$. Let $K = K_{p,\beta}$ denote the smallest subalgebra of $\mathcal{L}(E)$ which contains the operator $S = S_{\mathbb{R}+}$ of singular integration along the half line $\mathbb{R}^+$. For $0 < \delta < \frac{1}{2}$ we let $\ell(\delta)$ denote the circular arc which joins the points -1 and 1, and is characterized by the following properties: for every point $z \in \ell(\delta)$, $z \neq \pm 1$, the angle $-\widehat{1z1}$ is equal to $2\pi\delta$, and as one moves along $\ell(\delta)$ from the point -1 to 1, the segment $[-1,1]$ remains on the left. For numbers $\delta \in (\frac{1}{2}, 1)$ we set $\ell(\delta) = -\ell(1 - \delta)$. Finally, we put $\ell(\frac{1}{2}) = [-1,1]$. It is known (see [50(9)]) that the spectrum of the operator S in E coincides with the arc $\ell((1+\beta)/p)$. Since the algebra K is generated by a single element S, the following result holds:

THEOREM 13.5. *The set of maximal ideals of K is homeomorphic to the arc $\ell = \ell((1+\beta)/p)$. If M_z designates the maximal ideal corresponding to the point $z \in \ell$, then the Gelfand transformation is written: $S(M_z) = z$.*

$\square$

This theorem can be significantly sharpened.

THEOREM 13.6. $\mathcal{K}_{p,\beta}$ is an algebra with a symmetric involution $A \to \bar{A}$. In particular

$$\bar{S} = [cos(2\pi\gamma)S - i\ sin(2\pi\gamma)I][cos(2\pi\gamma)I - i\ sin(2\pi\gamma)S]^{-1}\ , \qquad (13.5)$$

where $\gamma = (1 + \beta)/p$. If the function $\mathcal{A}(z) = A(M_z)$ is the Gelfand transform of the operator A, then for $p = 2$

$$\|A\| = \max_{z \in \ell(\gamma)} |\mathcal{A}(z)|\ , \quad \gamma = (1 + \beta)/p \qquad (13.6)$$

whereas for $p \neq 2$

$$\max_{z \in \ell(\gamma)} |\mathcal{A}(z)| \leq \|A\| \leq c\ \max \left\{ \max_{z \in \ell(\gamma)} |\mathcal{A}(z)|, \max_{z \in \ell(\gamma)} \left|(1 - z^2)\ell n \frac{1 - z}{1 + z} \frac{d\mathcal{A}(z)}{dz}\right| \right\}, \qquad (13.7)$$

where the constant c depends only on B and β (the upper estimate in (13.7) is meaningful whenever the derivative $d\mathcal{A}(z)/dz$ exists).

PROOF. The operator B defined by the formula $(B\varphi)(t) = \varphi(e^t)e^{\gamma t}$, $t \in (-\infty, \infty)$, maps the space $L_p(\mathbb{R}^+, t^\beta)$ isometrically onto $L_p(\mathbb{R})$. It is readily verified that the operator $\tilde{S} = BSB^{-1}$ acts according to the rule

$$(\tilde{S}\psi)(t) = \frac{1}{\pi i} \int_{-\infty}^{\infty} e^{\gamma(t-s)}(1 - e^{t-s})^{-1}\psi(s)ds\ .$$

Hence, the algebra $\mathcal{K}_{p,\beta}$ generated by the operator S is isometrically isomorphic to a subalgebra of the convolution algebra. Let $\pi i\hat{S}(\xi)$ denote the Fourier transform of the function $e^{\gamma t}(1 - e^t)^{-1}$. A straightforward calculation gives

$$\hat{S}(\xi) = 1 + 2[exp(2\pi(\xi + i\gamma)) - 1]^{-1}\ , \quad -\infty < \xi < \infty\ .$$

The range of the function $\hat{S}(\xi)$ coincides with the arc $\ell(\gamma)$. Setting $z = \hat{S}(\xi)$ and denoting the Fourier transformation by F, we have

$$\mathcal{A}(z) = (FBAB^{-1}F^{-1})(\xi)$$

for every operator $A \in \mathcal{K}$. In particular, this yields equality (13.6) for $p = 2$. For $p \neq 2$ the estimate from below for $\|A\|$ follows from Theorem 13.5. The upper estimate is obtained with the help of a theorem of S.G. Mikhlin on multipliers (see [59]), according to which

$$\|BA^{-1}B\| \leq c_p \max \left\{ \max_{\xi \in \mathbb{R}} |\mathcal{A}(\hat{S}(\xi))|, \max_{\xi \in \mathbb{R}} \left|\xi \frac{d\mathcal{A}(\hat{S}(\xi))}{d\xi}\right| \right\}\ ,$$

in which the constant c_p depends only on p.

Finally, it is readily checked that $\bar{S}(z) = \overline{S(z)}$.

□

<u>Remark 13.1.</u> Let us write the functionals on $L_2(\mathbb{R}^+, t^\beta)$ in the form

$$f(\varphi) = \int_0^\infty \varphi(t)\overline{f(t)}t^\beta dt .$$

Then $S^* = t^{-\beta}St^\beta$, and it is easily seen that $FBS^*B^{-1}F^{-1} = FB\bar{S}B^{-1}F^{-1}$. Consequently, for $p = 2$, $\bar{S} = S^*$.

COROLLARY 13.3.   *Suppose the function f is differentiable at every point $z \in \ell(\gamma)\backslash\{-1,1\}$. If there exists a sequence of polynomials p_n such that*

$$\max_{z \in \ell(\gamma)} |p_n(z) - f(z)| \longrightarrow 0$$

and

$$\max_{z \in \ell(\gamma)} |(1 - z^2)\ell n\frac{1-z}{1+z}(p_n'(z) - f'(z))| \longrightarrow 0$$

as $n\rangle\infty$, then $f(S) \in K_{p,\beta}$.

□

More generally, we have

COROLLARY 13.4.   *Let $A_0 \in K_{p,\beta}$ and let $\varphi(z)$ designate the Gelfand transform of the operator A_0. Suppose that the function h is differentiable at every point $z \in \ell(\gamma)\backslash\{-1,1\}$. If there exists a sequence of polynomials p_n such that*

$$\max_{z \in \ell(\gamma)} |p_n(\varphi(z)) - h(z)|\rangle 0$$

and

$$\max_{z \in \ell(\gamma)} |(1 - z^2)\ell n\frac{1-z}{1+z}\frac{d}{dz}(p_n(\varphi(z)) - h(z))| \longrightarrow 0$$

as $n\rangle\infty$, then $h(A_0) \in K_{p,\beta}$.

□

We have thus shown that $K_{p,\beta} \in \pi_1$ and the family of multiplicative functionals $f_z(A) = A(z)$ is symmetric. By Theorem 13.3, if $A \in K_{p,\beta}$ then

$$A \in GL(E) \iff A \in GK_{p,\beta} \iff (f_z(A) = 0 \ \forall z \in \ell(\gamma)) \ .$$

THEOREM 13.7. *Let* $\omega = exp(\pi i \alpha)$*, where* $\alpha \in C$*. If* $-1 < Re \ \alpha < 1$*, then the operator* N_ω *defined by the rule*

$$(N_\omega \varphi)(x) = \frac{1}{\pi i} \int_0^\infty \frac{\varphi(y)\,dy}{y + \omega x} \ , \quad x \in \mathbb{R}^+ \ , \tag{13.8}$$

belongs to $K_{p,\beta}$ *and*

$$f_z(N_\omega) = (z - 1)^{(1+\alpha)/2}(z + 1)^{(1-\alpha)/2} \ , \quad z \in \ell(\gamma) \tag{13.9}$$

(the branch of this function is chosen so that for $z = -i \ cot(\pi\gamma)$ *it takes the value* $-i \ exp(-\pi i \gamma \alpha)/sin(\pi\gamma)$*).*

PROOF. It is readily verified that

$$(\pi i B N_\omega B^{-1})\varphi = [e^{\gamma t}(1 + \omega e^t)^{-1}] * \varphi \ ,$$

(where $*$ stands for convolution), which implies that

$$\begin{aligned} f_z(N_\omega) &= \frac{1}{\pi i} \int_{-\infty}^\infty \frac{e^{\gamma t - i\xi t}}{1 + \omega e^t}\,dt \\ &= \frac{-i\omega^{i\xi - \gamma}}{sin((\gamma - i\xi)\pi)} = (z - 1)^{(1+\alpha)/2}(z + 1)^{(1-\alpha)/2} \ . \end{aligned}$$

Let us show that the function $h(z) = f_z(N_\omega)$ satisfies the condition of Corollary 13.3 or 13.4. Suppose first that $|\gamma - \frac{1}{2}| \leq \frac{1}{4}$, in which case $|z| \leq 1$. Then for every δ with $Re \ \delta > 0$, the function $(1 \pm z)^\delta$ satisfies the condition of Corollary 13.3, in which for the polynomials $p_n(z)$ one can take the partial sums of its Taylor series. If $|\gamma - \frac{1}{2}| > \frac{1}{4}$, then the function $f_z(N_\omega) = z(1 - z^{-1})^{(1-\alpha)/2}(1 - z^{-1})^{(1+\alpha)/2}$ satisfies the condition of Corollary 13.4. The role of the operator A_0 is played by S^{-1}. The invertibility of S is guaranteed by the assumption that $\gamma \neq \frac{1}{2}$.

$\square$

14. SYMBOLS

Let $\mathcal{K}$ be a subalgebra of the algebra $\mathcal{L}(E)$ of all bounded linear operators in the Banach space E.

Definition 14.1. We say that $\mathcal{K}$ is an *algebra with symbol (semisymbol)* if there exists a family $\{\gamma_M\}_{M \in \alpha}$ of multiplicative functionals on $\mathcal{K}$ such that for any operator $A \in \mathcal{K}$

$$A \in \Phi(E) \Longleftrightarrow \gamma_M(A) \neq 0 \ \ \forall \, M \in \alpha$$

(respectively

$$A \in \Phi(E) \Longleftarrow \gamma_M(A) \neq 0 \ \ \forall \, M \in \alpha) \, .$$

We denote the class of algebras $\mathcal{K}$ with symbol (semisymbol) by σ_1 (respectively, σ_1^0). We shall assume, as a rule, that $\mathcal{T}(E) \subset \mathcal{K}$. In this case we let $\hat{\mathcal{K}}$ denote the quotient algebra $\mathcal{K}/\mathcal{T}(E)$, and we use $\hat{A}$ to designate the element of $\mathcal{K}$ containing the operator A. If $\mathcal{T}(E) \not\subset \mathcal{K}$, we put

$$\hat{\mathcal{K}} = \{\hat{A} \in \widehat{\mathcal{L}(E)} : \ \hat{A} \cap \mathcal{K} \neq \emptyset\} \, .$$

THEOREM 14.1. *Let $\mathcal{K} \subset \mathcal{L}(E)$, $\hat{\mathcal{K}} \in \pi_1$, and let $\{f_M\}_{M \in \alpha}$ be a sufficient family of multiplicative functionals. Then the rule*

$$\gamma_M(A) = f_M(\hat{A})$$

defines a semisymbol on the algebra $\mathcal{K}$. In order for $\{\gamma_M\}_{M \in \alpha}$ to be a symbol it is necessary and sufficient that $\hat{K}$ be inverse-closed in $\hat{\mathcal{L}} = \widehat{\mathcal{L}(E)}$.

PROOF. It is known that $A \in \Phi(E)$ if and only if $\hat{A} \in G\widehat{\mathcal{L}(E)}$ (see [50(9)]). Therefore

$$\begin{aligned} (\gamma_M(A) \neq 0 \ \ \forall \, M \in \alpha) &\Longleftrightarrow (f_M(\hat{A}) \neq 0 \ \ \forall \, M \in \alpha) \\ &\Longleftrightarrow \hat{A} \in G\hat{\mathcal{K}} \Longrightarrow \hat{A} \in G\widehat{\mathcal{L}(E)} \Longleftrightarrow A \in \Phi(E) \, , \end{aligned} \tag{14.3}$$

which proves that $\{\gamma_M\}_{M \in \alpha}$ is a semisymbol. If $\mathcal{K}$ is an inverse-closed subalgebra in $\widehat{\mathcal{L}(E)}$, then in (14.3) the converse of the implication $\Longrightarrow$ is also true, and hence the family $\{\gamma_M\}_{M \in \alpha}$ defines a symbol on $\mathcal{K}$.

If the algebra $\hat{\mathcal{K}}$ is not inverse-closed, then there is a Φ-operator $A \in \mathcal{K}$ such that $\hat{A} \notin G\hat{\mathcal{K}}$. Consequently, one can find a functional f_M such that $f_M(\hat{A}) = 0$, i.e. $\gamma_M(A) = 0$. Hence, the semisymbol $\{\gamma_M\}$ is not a symbol.

$\square$

Let us give an example of an algebra $\mathcal{K}$ for which $\hat{\mathcal{K}} \in \pi_1$, but the family of multiplicative functionals $\{\gamma_M\}$ defined by the formula $\gamma_M(A) = f_M(\hat{A})$, where f_M runs through a sufficient family of multiplicative functionals on $\hat{\mathcal{K}}$, does not define a symbol on $\mathcal{K}$.

Example 14.1. Let Γ_0 denote the unit circle, $E = L_2(\Gamma_0)$, and $C_+(\Gamma_0)$ the set of all functions holomorphic in the disc $\mathbb{D} = \{z \in \mathbb{C} : |z| < 1\}$ and continuous in $\overline{\mathbb{D}}$. We take for $\mathcal{K}$ the set of all operators of the form $H + T$, where $(H\varphi)(t) = h(t)\varphi(t)$ (with $h \in C_+(\Gamma_0)$) and $T \in \mathcal{T}(E)$. The space of maximal ideals of $\mathcal{K}$ is homeomorphic to the closed disc $\overline{\mathbb{D}}$. The multiplicative functionals $f_z(\widehat{H + T}) = h(z)$, $z \in \mathbb{D}$, form a sufficient family $\{f_z\}_{z \in \mathbb{D}}$. However, the semisymbol $\gamma_z(H + T) = h(z)$ is not a symbol: the operator H_0 of multiplication by the independent variable, $(H_0\varphi)(t) = t\varphi(t)$, is a Φ-operator in $L_2(\Gamma_0)$, but for $z_0 = 0$ we have $\gamma_{z_0}(H_0) = 0$.

In this example the semisymbol $\{\gamma_z\}_{z \in \overline{\mathbb{D}}}$ contains redundant functionals. One can extract from it the subfamily $\{\gamma_z\}_{z \in \Gamma_0}$, which is a symbol: $H + T \in \Phi(E) \iff h(t) \neq 0 \; \forall \, t \in \Gamma_0$. It turns out that a similar assertion is valid for every algebra $\mathcal{K}$ for which the quotient algebra $\hat{\mathcal{K}}$ is commutative.

THEOREM 14.2. *Let* $\mathcal{T}(E) \subset \mathcal{K} \subset \mathcal{L}(E)$ *and suppose that the quotient algebra* $\hat{\mathcal{K}} = \mathcal{K}/\mathcal{T}(E)$ *is commutative. Let* $\mathcal{M}$ *be the set of all maximal ideals of* $\hat{\mathcal{K}}$ *and* $\{f_M\}_{M \in \mathcal{M}}$ *the corresponding set of multiplicative functionals. Then there exists a subset* $\mathcal{M}_0$ *of* $\mathcal{M}$ *such that the family of functionals* $\gamma_M(A) = f_M(\hat{A})$, $M \in \mathcal{M}_0$, *is a symbol on* $\mathcal{K}$.

If the family $\{f_M\}_{M \in \mathcal{M}}$ *is symmetric, then one can take* $\mathcal{M}_0 = \mathcal{M}$.

PROOF. Every commutative subalgebra $\hat{K}$ of $\widehat{\mathcal{L}(E)}$ can be imbedded in a commutative inverse-closed subalgebra $\mathcal{K}^\bullet$ of $\widehat{\mathcal{L}(E)}$: for example, one can take for $\mathcal{K}^\bullet$ the maximal commutative subalgebra $\mathcal{K}_{\max}$ of $\widehat{\mathcal{L}(E)}$ which contains $\hat{K}$. If $\hat{A} \in \mathcal{K}_{\max} \cap G\widehat{\mathcal{L}(E)}$, then $\hat{A}^{-1}\hat{B} = \hat{B}\hat{A}^{-1}$ for all $\hat{B} \in \mathcal{K}_{\max}$, and hence $\hat{A}^{-1} \in \mathcal{K}_{\max}$, which says that $\mathcal{K}_{\max}$ is an inverse-closed subalgebra in $\widehat{\mathcal{L}(E)}$. We let $\mathcal{M}_0$ denote the subset of $\mathcal{M}$ consisting of all maximal ideals M for which f_M extends to a multiplicative functional on $\mathcal{K}^\bullet$. Since every multiplicative functional on $\mathcal{K}^\bullet$ is the extension of its restriction to $\hat{K}$, the set $\{f_M\}_{M \in \mathcal{M}_0}$ is actually equal to the set of all multiplicative functionals on $\mathcal{K}^\bullet$.

We next show that the family $\{\gamma_M\}_{M \in \mathcal{M}_0}$ is a symbol on $\mathcal{K}$. Let $A \in \mathcal{K}$. Then

$$A \in \Phi(E) \Longleftrightarrow \hat{A} \in G\widehat{\mathcal{L}(E)} \Longleftrightarrow \hat{A} \in G\mathcal{K}^\bullet$$

$$\Longleftrightarrow (f_M(\hat{A}) \neq 0 \ \ \forall M \in \mathcal{M}_0) \Longleftrightarrow (\gamma_M(A) \neq 0 \ \ \forall M \in \mathcal{M}_0) .$$

If the family $\{f_M\}_{M \in \mathcal{M}}$ is symmetric, then by Theorem 13.3, $\mathcal{K}$ is inverse-closed in $\widehat{\mathcal{L}(E)}$, and hence $\{\gamma_M\}_{M \in \mathcal{M}}$ is a symbol on $\mathcal{K}$.

□

COROLLARY 14.1. *If the quotient algebra $\hat{K}$ is commutative, then $\mathcal{K} \in \sigma_1$.*

□

Remark 14.1. In Example 14.1, algebra $\hat{K}$ is not inverse-closed in $\widehat{\mathcal{L}(E)}$. The maximal commutative subalgebra $\mathcal{K}_{\max}$ is isometrically isomorphic to $L_\infty(\Gamma_0)$, but it is not obligatory to take $\mathcal{K}^\bullet = \mathcal{K}_{\max}$. Here it is more convenient to take for $\mathcal{K}^\bullet$ the set of all elements $\widehat{H + T}$ with $h \in C(\Gamma_0)$. For this choice $\mathcal{K}^\bullet$ is isometrically isomorphic to $C(\Gamma_0)$ and $\mathcal{M}_0$ is homeomorphic to the unit circle Γ_0.

Remark 14.2. In Theorem 14.2 the condition that $\hat{K}$ be commutative cannot be replaced by $\hat{K} \in \pi_1$.

We next give an example of a Banach algebra $\mathcal{K}$ such that $\hat{K} \in \pi_1$ but $\mathcal{K} \notin \sigma_1$.

Example 14.2. Let $E = \ell_1 \times \ell_2$. The formulas

$$H_1 x = (0, \xi_1, 0, \xi_2, \ldots) \quad \text{and} \quad H_2 x = (\xi_2, \xi_4, \xi_6, \ldots)$$

define operators H_1 and H_2 in ℓ_2, such that $H_2 H_1 = I$ and $dim\, Ker\, H_2 = \infty$. We let $\mathcal{K}_m$ $(m = 1, 2)$ denote the smallest Banach subalgebra (with identity) of $\mathcal{L}(\ell_2)$ which contains

H_m and all compact operators. Let $\mathcal{K}_3$ designate the subalgebra of $\mathcal{L}(\ell_2)$ generated by the operators from $\mathcal{K}_1$ and $\mathcal{K}_2$. Now for $\mathcal{K}$ we take the subalgebra of $\mathcal{L}(E)$ consisting of all operators $A = [B_{jk}]^2_{j,k=1}$ with $B_{11} \in \mathcal{K}_1$, $B_{22} \in \mathcal{K}_2$, $B_{12} \in \mathcal{K}_3$, and $B_{21} \in T(\ell_2)$. In this example $\hat{\mathcal{K}}$ consists of all elements of the form $[\hat{B}_{jk}]^2_{j,k=1}$ where $\hat{B}_{21} = 0$ and $\hat{B}_{11}$ ($\hat{B}_{22}$) belongs to the commutative algebra $\hat{\mathcal{K}}_1$ (respectively, $\hat{\mathcal{K}}_2$). The set of elements $[\hat{B}_{jk}] \in \hat{\mathcal{K}}$ for which $\hat{B}_{jk} = 0$ if $j \geq k$ is the radical $\mathcal{R}(\hat{\mathcal{K}})$. The quotient algebra $\hat{\mathcal{K}}/\mathcal{R}(\hat{\mathcal{K}})$ is commutative. Hence, by Theorem 5.1, $\hat{\mathcal{K}} \in \pi_1$. We show that $\mathcal{K} \notin \sigma_1$ by reductio ad absurdum. Suppose that $\mathcal{K} \in \sigma_1$ and let $\{\gamma_M\}_{M\in\alpha}$ be a family of multiplicative functionals which defines a symbol on $\mathcal{K}$. Then

$$\gamma_M\left(\begin{bmatrix} B_{11} & 0 \\ 0 & B_{22} \end{bmatrix}\right)$$

$$= \gamma_M\left(\begin{bmatrix} B_{11} & B_{12} \\ 0 & B_{22} \end{bmatrix}\begin{bmatrix} I & 0 \\ 0 & 0 \end{bmatrix} + \begin{bmatrix} 0 & 0 \\ 0 & I \end{bmatrix}\begin{bmatrix} B_{11} & B_{12} \\ 0 & B_{22} \end{bmatrix}\right)$$

$$= \gamma_M\left(\begin{bmatrix} I & 0 \\ 0 & 0 \end{bmatrix}\begin{bmatrix} B_{11} & B_{12} \\ 0 & B_{22} \end{bmatrix} + \begin{bmatrix} 0 & 0 \\ 0 & I \end{bmatrix}\begin{bmatrix} B_{11} & B_{12} \\ 0 & B_{22} \end{bmatrix}\right) \tag{14.4}$$

$$= \gamma_M\left(\begin{bmatrix} B_{11} & B_{12} \\ 0 & B_{22} \end{bmatrix}\right)$$

(here we used the fact that the functional γ_M is multiplicative). This shows that for arbitrary B_{jk} in the appropriate algebras

$$\begin{bmatrix} B_{11} & 0 \\ 0 & B_{22} \end{bmatrix} \in \Phi(E) \iff \begin{bmatrix} B_{11} & B_{12} \\ 0 & B_{22} \end{bmatrix} \in \Phi(E).$$

But

$$\begin{bmatrix} H_2 & -I \\ I-H_1H_2 & H_1 \end{bmatrix}\begin{bmatrix} H_1 & I \\ 0 & H_2 \end{bmatrix} = \begin{bmatrix} H_1 & I \\ 0 & H_2 \end{bmatrix}\begin{bmatrix} H_2 & -I \\ I-H_1H_2 & H_1 \end{bmatrix} = \begin{bmatrix} I & 0 \\ 0 & I \end{bmatrix},$$

and hence the operator $\begin{bmatrix} H_1 & I \\ 0 & H_2 \end{bmatrix}$ is invertible in E. On the other hand, the operator $\begin{bmatrix} H_1 & I \\ 0 & H_2 \end{bmatrix}$ is not Fredholm in E because $dim\,Ker\,H_2 = \infty$.

We have thus shown that $\hat{\mathcal{K}} \in \pi_1$ but $\mathcal{K} \notin \sigma_1$.

Thus far we have been concerned only with conditions sufficient for the existence of a symbol. To discuss necessary conditions we need the following definition:

<u>Definition 14.2.</u> We say that the operator $T \in \mathcal{L}(E)$ is a Φ-*admissible pertur-*
bation for the algebra $\mathcal{K} \subset \mathcal{L}(E)$ if

$$A \in \Phi(E) \cap \mathcal{K} \Longrightarrow A + T \in \Phi(E) .$$

LEMMA 14.1. *Let* $T(E) \subset \mathcal{K} \subset \mathcal{L}(E)$. *Then the set* Σ *of all* Φ-*admissible*
perturbations of algebra $\mathcal{K}$ *which themselves belong to* $\mathcal{K}$ *is a closed two-sided ideal in* $\mathcal{K}$.
If, in particular, $\hat{\mathcal{K}}$ *is a semisimple inverse-closed subalgebra of* $\widehat{\mathcal{L}(E)}$, *then* $\Sigma = T(E)$.

PROOF. The fact that the set Σ is linear is obvious. Let $A \in \mathcal{K}$, $T \in \Sigma$, $B \in \mathcal{K} \cap \Phi(E)$, and λ a regular point of the element $A \in \mathcal{K}$. Then $B + AT = (A - \lambda I)[(A - \lambda I)^{-1}B + T] + \lambda T \in \Phi(E)$ and hence $AT \in \Sigma$. In the same way one proves that $TA \in \Sigma$. If $T_n \in \Sigma$ and $\|T_n - T_0\| \to 0$, then $B + T_0 - T_n \in \Phi(E)$ for sufficiently large n, and hence $T_0 \in \Sigma$. We have thus shown that Σ is a closed two-sided ideal in $\mathcal{K}$.

Now let $T \in \Sigma$. Then $I + CT \in \Phi(E)$ for every operator $C \in \mathcal{K}$, and hence $\hat{I} + \hat{C}\hat{I} \in G\widehat{\mathcal{L}(E)}$. If $\hat{\mathcal{K}}$ is inverse-closed in $\widehat{\mathcal{L}(E)}$, then $\hat{I} + \hat{C}\hat{I} \in G\hat{\mathcal{K}}$, and hence $\hat{T} \in \mathcal{R}(\hat{\mathcal{K}})$. If, in addition, $\hat{\mathcal{K}}$ is semisimple, then $\hat{T} = 0$, i.e., $T \in T(E)$.

$\square$

THEOREM 14.3. *Let* $\mathcal{K} \subset \mathcal{L}(E)$. *If* $\mathcal{K} \in \sigma_1$, *then for arbitrary* $A, B \in \mathcal{K}$
the operator $T = AB - BA$ *is a* Φ-*admissible perturbation for the algebra* $\mathcal{K}$. *If* $\hat{\mathcal{K}}$ *is a*
semisimple inverse-closed subalgebra in $\widehat{\mathcal{L}(E)}$, *then* T *is compact.*

PROOF. Suppose that $\mathcal{K} \in \sigma_1$. Let $A, B, C \in \mathcal{K}$, and set $T = AB - BA$. Since $\gamma_M(C + T) = \gamma_M(C)$ for every M, $C + T \in \Phi(E)$ if and only if $C \in \Phi(E)$, i.e., T is a Φ-admissible perturbation for $\mathcal{K}$. If $\hat{\mathcal{K}}$ is a semisimple inverse-closed subalgebra in $\widehat{\mathcal{L}(E)}$ then, by Lemma 14.1, $T \in T(E)$.

$\square$

We give an example of algebra $\mathcal{K} \in \sigma_1$ for which $T = AB - BA \notin T(E)$. Let E be a Banach space, let $\mathcal{J}$ be the set of operators $[A_{jk}]_{j,k=1}^3 \in \mathcal{L}(E^3)$ such that $A_{jk} =$ for $j \geq k$, and let $\mathcal{K}$ be the set of all operators of the form $\lambda I + T + R$ with $\lambda \in \mathbb{C}$,

$R \in \mathcal{J}$, and $T \in \mathcal{T}(E^3)$. It is readily checked that $\lambda I + T + R \in \Phi(E^3) \iff \lambda \neq 0$. In this example the single homomorphism $\gamma(\lambda I + T + R) = \lambda$ defines a symbol, and consequently $\mathcal{K} \in \sigma_1$. However, for the operators A and B defined by the rules $A(x, y, z) = (y, 0, 0)$ and $B(x, y, z) = (0, z, 0)$, the commutator $AB - BA$ does not belong to $\mathcal{T}(E)$.

COROLLARY 14.2.   *Let E be a Hilbert space and let $\mathcal{K}$ be a self-adjoint subalgebra of $\mathcal{L}(E)$. Then $\mathcal{K} \in \sigma_1$ if and only if the quotient algebra $\hat{\mathcal{K}} = \mathcal{K}/\mathcal{T}(E)$ is commutative.*

PROOF.   In this case $\hat{\mathcal{K}}$ is a self-adjoint inverse-closed subalgebra in $\widehat{\mathcal{L}(E)}$ and Theorems 14.1 and 14.3 apply.

□

Example 14.3.   Let $\mathbb{R}_+^m = \{t = (t_1, \ldots, t_m) \in \mathbb{R}^m : t_1 \geq 0\}$, $E = L_2(\mathbb{R}_+^m)$, and $\mathcal{K}$ the subalgebra of $\mathcal{L}(E)$ generated by the Wiener-Hopf operators

$$(A\varphi)(t) = \int_{\mathbb{R}_+^m} k(t - s)\varphi(s)ds \,, \quad t \in \mathbb{R}_+^m \,, \tag{14.5}$$

where $k \in L_1(\mathbb{R}^m)$. For $m = 1$, $\mathcal{K}$ consists of the operators $B = \lambda I + A + T$ with $\lambda \in \mathbb{C}$, $T \in \mathcal{T}(E)$, and A an operator of the type (14.5). In [37] it is shown that $B \in \Phi(E)$ if and only if $\lambda + \tilde{k}(t) \neq 0$ on $\mathbb{R} \cup \{\infty\}$, where $\tilde{k}$ designates the Fourier transform of the function k. The formula $f_t(B) = \lambda + \tilde{k}(t)$ defines a family of multiplicative functionals which is a symbol on $\mathcal{K}$. Therefore, if $m = 1$, then $\mathcal{K} \in \sigma_1$.

Let us show that for $m > 1$, $\mathcal{K} \notin \sigma_1$. To this end it suffices, in view of Corollary 14.2, to exhibit two operators $A_1, A_2 \in \mathcal{K}$ for which $T = A_1 A_2 - A_2 A_1 \notin \mathcal{T}(E)$. Let $c_j \in L_1(\mathbb{R})$, $j = 1, 2$, $b \in L_1(\mathbb{R}^{m-1})$, and set

$$(C_j \varphi)(t_1) = \int_0^\infty c_j(t_1 - s_1)\varphi(s_1)ds_1$$

and

$$(B\psi)(\tau) = \int_{\mathbb{R}^{m-1}} b(\tau - \sigma)\psi(\sigma)d\sigma \,,$$

where we put $t = (t_1, t_2, \ldots, t_n) = (t_1, \tau)$ and $s = (s_1, s_2, \ldots, s_n) = (s_1, \sigma)$.   One can choose the functions c_j, $j = 1, 2$, and b so that the operator $T_0 = C_1 C_2 - C_2 C_1$ in

$L_2(\mathbb{R}_+^1)$ will be different from zero and $B \notin T(L_2(\mathbb{R}_+^{m-1}))$. In this case one can find a $\psi \in L_2(\mathbb{R}_+^{m-1})$ and a sequence $\varphi_\ell \in L_2(\mathbb{R}_+^1)$ such that $\|\varphi_\ell\| = \|\psi\| = 1$ and the sequence of functions $g_\ell(t) = (T_0 \varphi_\ell)(t_1)(B\psi)(\tau)$ is not compact in $L_2(\mathbb{R}_+^m)$. We let A_j denote the operators of the form (15.4) with kernels $k_j(t) = c_j(t_1)b(\tau)$ $(j = 1, 2, \; t = (t_1, \tau) \in \mathbb{R}^m)$ and put $h_\ell(t) = \varphi_\ell(t_1)\psi(\tau)$. Then it is readily verified that $(A_1 A_2 - A_2 A_1)h_\ell = g_\ell$, from which it follows that $T = A_1 A_2 - A_2 A_1 \notin T(E)$.

15. A LOCAL PRINCIPLE

In this section we discuss one of the variants of the local principle which, thanks to its generality and simplicity, finds numerous applications.

Let $\mathcal{K}$ be a Banach algebra with identity e. We call a subset M of $\mathcal{K}$ a *localizing class* if $0 \notin M$ and for every pair $a_1, a_2 \in \mathcal{K}$ there is an element $a \in \mathcal{K}$ such that $a_j a = a a_j = a, \; j = 1, 2$.

Two elements $x, y \in \mathcal{K}$ are said to be *left M-equivalent* if $\inf_{a \in M} \|(x - y)a\| = 0$. The *right M-equivalence* is defined in the same manner. If the elements $x, y, \in \mathcal{K}$ are both left and right equivalent, then we say that they are *M-equivalent*.

Finally, we say that $x \in \mathcal{K}$ is *left (right) M-invertible* if there are elements $z \in \mathcal{K}$ and $a \in M$ such that $zxa = a$ (respectively, $axz = a$).

LEMMA 15.1.   *Suppose that M is a localizing class and $x, y \in \mathcal{K}$ are left (respectively, right) M-equivalent. If x is left (respectively, right) M-invertible, then y is left (respectively, right) M-invertible.*

PROOF. Let x be left M-invertible, i.e., there are elements $z \in \mathcal{K}$ and $a_1 \in M$ such that $zxa_1 = a_1$. Since x and y are left M-equivalent, there exists an $a_2 \in M$ such that $\|(x - y)a_2\| < \|z\|^{-1}$. Pick $a \in M$ so that $a_1 a = a_2 a = a$. Then $zya = zxa - ua$, where $u = z(x - y)a_2$. Taking into account that $zxa = a$, we see that $zya = (e - u)a$. Since $\|u\| < 1$, it follows that $e - u \in G\mathcal{K}$. Therefore, $z_1 ya = a$, where $z_1 = (e - u)^{-1}z$, i.e., y is left M-invertible.

□

Let $\{M_s\}_{s\in S}$ be a system of localizing classes. We say that $\{M_s\}_{s\in S}$ is a *covering system* if from every family $\{a_s\}_{s\in S}$, $a_s \in M_s$, one can extract a finite family whose sum is an invertible element.

LEMMA 15.2. *Let $\{M_s\}_{s\in S}$ be a covering system of localizing classes. Suppose that $x \in \mathcal{K}$ commutes with all the elements from $\bigcup_{s\in S} M_s$. Then x is left (right) invertible in $\mathcal{K}$ if and only if x is left (respectively, right) M_s-invertible for every $s \in S$.*

PROOF. The necessity of this condition is obvious. Let us prove its sufficiency. Suppose that x is left M_s-invertible for every $s \in S$. Then there exist elements $z_s \in \mathcal{K}$ and $a_s \in M_s$ such that $z_s x a_s = a_s$. Since $\{M_s\}$ is covering, we can extract from $\{a_s\}$ a finite subfamily $a_{s_1}, \ldots, a_{s_N}$ such that $\Sigma_{j=1}^N a_{s_j}$ is invertible. Let $u = \Sigma_{j=1}^N z_{s_j} a_{s_j}$. Then

$$ux = \sum_{j=1}^N z_{s_j} a_{s_j} x = \sum_{j=1}^N z_{s_j} x a_{s_j} = \sum_{j=1}^N a_{s_j} \ ,$$

and hence x is left invertible. Right invertibility is established in the same manner.

□

A straightforward consequence of these two lemmas is

THEOREM 15.1. *Let $\{M_s\}_{s\in S}$ be a covering system of localizing classes, and let x be left (right) M_s-equivalent to an element $y_s \in \mathcal{K}$ for every $s \in S$. Suppose that x commutes with all the elements from $\bigcup_{s\in S} M_s$. Then x is left (respectively, right) invertible if and only if y_s is left (respectively, right) M_s-invertible for every $s \in S$.*

□

We shall also need the following theorem:

THEOREM 15.2. *Let M_1 and M_2 be two localizing classes and let $v \in G\mathcal{K}$ be such that the map $x \to v^{-1}xv$ is a bijection of M_1 onto M_2. If $a \in \mathcal{K}$ and $v^{-1}av = a$, then a is left (right) M_1-invertible if and only if it is left (respectively, right) M_2-invertible.*
The proof is obvious.

□

We consider some examples.

Example 15.1. Let $\mathcal{K} = C(0,1)$ and $s \in [0,1]$. We let M_s denote the set of all nonnegative functions $x \in C(0,1)$ such that x is equal to 1 in some neighborhood of the point s. Then M_s is a localizing class. Two functions $x, y \in C(0,1)$ are M_s-equivalent if $x(s) = y(s)$. A function $x \in C(0,1)$ is M_s-invertible if $x(s) \neq 0$. The family $\{M_s\}_{s \in [0,1]}$ is a covering system.

Example 15.2. Let $E = L_p(\Gamma_0)$ $(1 < p < \infty)$ and $\mathcal{K} = \widehat{\mathcal{L}(E)}$. For each point $\tau \in \Gamma_0$ we set $M_\tau = \{\hat{H}\}$, where $(H\varphi)(t) = h(t)\varphi(t)$ with $h \in C(\Gamma_0)$ such that $h(t) \geq 0$ on Γ_0 and $h(t) = 1$ in some neighborhood of the point τ. It is readily verified that $\{M_\tau\}_{\tau \in \Gamma_0}$ is a covering system of localizing classes. Let S be the operator of singular integration along Γ_0, $\alpha, \beta \in \mathbb{C}$, and $A = \alpha I + \beta S$. Since $S^2 = I$ and $\hat{S} \neq \pm \hat{I}$,

$$\hat{A} \in G\hat{\mathcal{K}} \Longleftrightarrow \alpha^2 - \beta^2 \neq 0 .$$

Let $\tau, \zeta \in \Gamma_0$ and set $\varepsilon = \tau\zeta^{-1}$ and $(V\varphi)(t) = \varphi(\varepsilon t)$, $t \in \Gamma_0$. Then obviously $\hat{V}^{-1} M_\zeta \hat{V} = M_\tau$ and $V^{-1}SV = S$. By Theorem 15.2, $\hat{A}$ is M_τ-invertible for some point τ if and only if it is M_τ-invertible for all points τ, i.e., if and only if $\alpha^2 - \beta^2 \neq 0$.

Example 15.3. Let $\mathcal{K}$ be the algebra of Example 15.2. Let $\hat{B} \in \mathcal{K}$, where $B = aI + bS$ and $a, b \in C(\Gamma_0)$. It is readily checked that $\hat{B}$ is M_τ-equivalent to the element $\hat{B}_\tau$, where $B_\tau = a(\tau)I + b(\tau)S$, a singular integral operator with constant coefficients. Moreover, $\hat{B}\hat{H} = \hat{H}\hat{B}$ for every $\hat{H} \in M_\tau$. By Theorem 14.1,

$$B \in \Phi(E) \Longleftrightarrow \hat{B} \in G\hat{\mathcal{K}} \Longleftrightarrow (\hat{B}_\tau \in G_\tau \mathcal{K} \quad \forall \tau \in \Gamma_0) \Longleftrightarrow (a^2(\tau) - b^2(\tau) \neq 0 \quad \forall \tau \in \Gamma_0) ,$$

where $G_\tau \mathcal{K}$ designates the set of M_τ-invertible elements of algebra $\mathcal{K}$. Therefore, the algebra of all operators $A = aI + bS + T$ with $a, b \in C(\Gamma_0)$ and $T \in \mathcal{T}(E)$ possesses a symbol; the latter is defined by the family of multiplicative functionals $\gamma_t^+(A) = a(t) + b(t)$, $\gamma_t^-(A) = a(t) - b(t)$.

16. THE ALGEBRA GENERATED BY THE TOEPLITZ OPERATORS WITH PIECEWISE-CONTINUOUS COEFFICIENTS

Let $a \in L_\infty(\Gamma)$, let $a_i, i \in \mathbb{Z}$, be the Fourier coefficients of a, and let $T_0 = [a_{j-k}]_{j,k=0}^\infty$ denote the operator in ℓ_2 defined by the Toeplitz matrix built from the a_i's.

LEMMA 16.1. *Let $a \in L_\infty(\Gamma_0)$ and $b \in C(\Gamma_0)$. Then $R_1 = T_a T_b - T_{ab} \in \mathcal{T}(\ell_2)$* and $R_2 = T_b T_a - T_{ab} \in \mathcal{T}(\ell_2)$.

PROOF. It is readily verified that if b is a trigonometric polynomial, then R_1 and R_2 are finite-rank operators. Since $\|T_c\|_{\ell_2} = \|c\|_{L_\infty(\Gamma_0)}$ for every $c \in L_\infty(\Gamma_0)$ (see [50(9)]), the assertion of the lemma is established by approximating b uniformly on Γ_0 by trigonometric polynomials.

□

LEMMA 16.2. *Suppose that the functions $a, b \in PC(\Gamma_0)$ have no common points of discontinuity. Then $T_a T_b - T_{ab} \in \mathcal{T}(\ell_2)$*

PROOF. Let $t_1, \ldots, t_\ell$ $(t_{\ell+1}, \ldots, t_r)$ denote all the points of discontinuity of the function a (respectively, b). Let φ and ψ be continuous functions on Γ_0 with the properties: $\varphi(t_k) = 0$, $k = 1, \ldots, \ell$, $\psi(t_k) = 0$, $k = \ell + 1, \ldots, r$, and $\varphi(t) + \psi(t) \equiv 1$. It is readily checked that $a\varphi$ and $b\psi$ are continuous on Γ_0. Using 16.1, we have

$$T_{ab} = T_{a(\varphi+\psi)b} = T_{a\varphi b} + T_{a\psi b} = T_{a\varphi}T_b + T_a T_{b\psi} + T'$$

$$= (T_\varphi + T_\psi)T_a T_b + T'' = T_a T_b + T''$$

(with $T', T'' \in \mathcal{T}(\ell_2)$).   □

LEMMA 16.3. *If $a, b \in PC(\Gamma_0)$, then $T_a T_b - T_b T_a \in \mathcal{T}(\ell_2)$.*

PROOF. Let $t_1, \ldots, t_n$ denote all the points of discontinuity of the functions a and b. Let h be a function which is continuous at every point $t \in \Gamma_0 \backslash \{t_1, \ldots, t_n\}$ and satisfies the conditions $h(t_k + 0) = 1$, $h(t_k - 0) = 0$. Now let $\tilde{a}$ $(\tilde{b})$ be a continuous function such that $\tilde{a}(t_k) = a(t_k + 0) - a(t_k - 0)$ (respectively, $\tilde{b}(t_k) = b(t_k + 0) - b(t_k - 0)$). It is readily checked that the functions $\varphi = a - \tilde{a}h$ and $\psi = b - \tilde{b}h$ are continuous on Γ_0. Using Lemma 16.1, we have

$$T_a T_b - T_b T_a = (T_{\tilde{a}} T_h + T_\varphi)(T_{\tilde{b}} T_h + T_\psi)$$

$$- (T_{\tilde{b}} T_h + T_\psi)(T_{\tilde{a}} T_h + T_\varphi) + T' = T'' \ ,$$

(with $T', T'' \in \mathcal{T}(\ell_2)$).   □

We let $\mathcal{K}$ denote the smallest Banach subalgebra of $\mathcal{L}(\ell_2)$ which contains all the operators T_a with $a \in PC(\Gamma_0)$ and all compact operators. (It is not hard to show that the last requirement is superfluous, since $T(\ell_2)$ is already contained in the algebra generated by the Toeplitz operators T_a with continuous symbols a; see [10].)

Since $T_{\bar{a}} = T_a^*$, $\mathcal{K}$ is a C^*-subalgebra of $\mathcal{L}(\ell_2)$. By Lemma 16.3, $\hat{\mathcal{K}}$ is a commutative C^*-algebra. Let $\mathcal{M}$ denote the set of all maximal ideals of $\hat{\mathcal{K}}$. By the Gelfand-Naimark theorem, $\hat{\mathcal{K}}$ is isometrically isomorphic to the algebra of all continuous functions on the compact space $\mathcal{M}$. It follows from Theorem 14.2 that $\mathcal{K} \in \sigma_1$, and its symbol is defined by the family of multiplicative functionals $\gamma_M(A) = f_M(\hat{A})$, where f_M runs through the set of all multiplicative functionals on $\hat{\mathcal{K}}$.

Let us describe the compactum $\mathcal{M}$.

THEOREM 16.1.   *The quotient algebra $\hat{\mathcal{K}}$ is isometrically isomorphic to the algebra of all continuous functions on the cylinder $\Gamma_0 \times [0,1]$, endowed with the topology in which the neighborhoods of the points (ζ_0, μ), $\zeta_0 = exp(i\varphi_0)$, are defined as follows:*

$$
\left.
\begin{aligned}
U_{\varepsilon,\delta}(\zeta_0, 0) &= \{(\zeta,\mu): \ \varphi_0 - \delta < arg \ \zeta < \varphi_0; \ 0 \le \mu \le 1\} \cup \{(\zeta_0,\mu): \ 0 \le \mu < \varepsilon\}, \\
U_{\varepsilon,\delta}(\zeta_0, 1) &= \{(\zeta,\mu): \ \varphi_0 < arg \ \zeta < \varphi_0 + \delta; \ 0 \le \mu \le 1\} \cup \{(\zeta_0,\mu): \ \varepsilon < \mu \le 1\}, \\
U_{\delta_1,\delta_2}(\zeta_0, \mu_0) &= \{(\zeta_0,\mu): \ \mu_0 < \delta_1 < \mu < \mu_0 + \delta_2\}, \ \mu_0 \ne 0,1
\end{aligned}
\right\}
$$

(16.1)

with arbitrary $0 < \delta_1 < \mu_0$, $0 < \delta_2 < 1 - \mu_0$, and $0 < \varepsilon < 1$.

This isometric isomorphism takes the element $\hat{A} = \hat{T}_a$ into the function

$$
a(t,\mu) = a(t+0)\mu + a(t-0)(1-\mu), \quad (t,\mu) \in \Gamma_0 \times [0,1] \ .
$$

PROOF. Let (t_0, μ_0) be an arbitrary point of the cylinder $\Gamma_0 \times [0,1]$. We show that the functional defined on the generators $\hat{T}_a$ of $\hat{\mathcal{K}}$ by the rule

$$
f_{t_0,\mu_0}(\hat{T}_a) = a(t_0, \mu_0)
$$

(16.2)

extends to a multiplicative functional on the whole algebra $\hat{\mathcal{K}}$.

Let

$$
\hat{A} = \sum_{j=1}^{m} \prod_{k=1}^{j} \hat{T}_{a_{jk}} \ ,
$$

and

$$\hat{B} = \hat{A}\hat{A}^* - \lambda \hat{I} = (\sum \prod \hat{T}_{a_{jk}})(\sum \prod \hat{T}_{a_{jk}}^*) - \lambda \hat{I} \ .$$

Let $\Xi(\hat{B})$ be the linear dilation of the element $\hat{B}$ corresponding to the indicated decomposition (see Sec.1). By Corollary 1.1,

$$\hat{B} \in \widehat{G\mathcal{L}(\ell_2)} \Longleftrightarrow \Xi(\hat{B}) \in G\widehat{\mathcal{L}(\ell_2)}^{s \times s} \Longleftrightarrow \Xi(B) \in \Phi(\ell_2^s) \ .$$

But $\Xi(B)$ is the Toeplitz operator T_φ associated with the matrix-function $\varphi = \Xi(b)$, where $b = (\sum \prod a_{jk})(\sum \prod \bar{a}_{jk}) - \lambda$. It follows from Theorem 10.4 that

$$\Xi(B) \in \Phi(\ell_2^s) \Longleftrightarrow (det(\mu\varphi(1+0) + (1-\mu)\varphi(t-0)) \neq 0 \quad \text{on} \quad \Gamma_0 \times [0,1]) \ .$$

We set $\varphi(t,\mu) = \mu\varphi(t+0) + (1-\mu)\varphi(t-0)$. It follows from the definition of the linear dilation that

$$det \ \varphi(t,\mu) = |\sum \prod \ a_{jk}(t,\mu)|^2 - \lambda = |f_{t,\mu}(\hat{A})|^2 - \lambda \ .$$

Consequently, the spectrum of the element $\hat{A}\hat{A}^*$ in the algebra $\widehat{\mathcal{L}(\ell_2)}$ (and hence in $\hat{\mathcal{K}}$) coincides with the range of the function $|f_{t,\mu}(\hat{A})|^2$ for $(t,\mu) \in \Gamma_0 \times [0,1]$. We thus obtain the equality

$$\max_{(t,\mu) \in \Gamma_0 \times [0,1]} |f_{t,\mu}(\hat{A})| = \|\hat{A}\| \ , \tag{16.3}$$

from which it follows that the functional $f_{t,\mu}(\hat{A})$ is independent of the way in which $\hat{A}$ is represented as a sum of products of $\hat{T}_{a_{jk}}$. This in turn implies that the functional $f_{t,\mu}$ is multiplicative and bounded on the dense subset $\hat{\mathcal{K}}_0 = \{\sum \prod \hat{T}_{a_{jk}}\}$ of $\hat{\mathcal{K}}$, and hence on the entire $\hat{\mathcal{K}}$. The functionals $f_{t,\mu}$ with $(t,\mu) \in \Gamma_0 \times [0,1]$ define all the maximal ideals of the algebra $\hat{\mathcal{K}}$. An analogous statement was proved in Theorem 16.4 for the spaces h_p. Since $h_2 = \ell_2$, we shall not discuss this point further.

To complete the proof it remains to show that the topology defined by the neighborhoods (16.1) coincides with the Gelfand topology on $\mathcal{M}$. It is readily verified that these neighborhoods define on the cylinder $\mathcal{M}$ a Hausdorff compact topology for which the functions $f_{t,\mu}(\hat{A})$ are continuous. By Theorem 1′ of [18, Sec.5], this topology coincides with the Gelfand topology on $\mathcal{M}$.   □

COROLLARY 16.1. *The family of multiplicative functionals* $\{\gamma_M\}$ *($M \in \mathcal{M}$) on the algebra* $\mathcal{K}$, *given on generators* T_a *by the formula* $\gamma_{t,\mu}(T_a) = a(t,\mu)$, *defines a symbol on* $\mathcal{K}$.

$\square$

Let $A = \sum \prod T_{a_{jk}}$ and $\mathcal{A}(t,\mu) = f_{t,\mu}(\hat{A}) = \sum \prod a_{jk}(t,\mu)$. If $\mathcal{A}(t,\mu) \neq 0$ on $\Gamma_0 \times [0,1]$, then

$$Ind \; A = Ind \; \Xi(A) = -ind \; det \; \Xi(\mathcal{A}(t,\mu)) = -ind \; \mathcal{A}(t,\mu)$$

(see (10.3)). If $A \in \mathcal{K}$, then there is a sequence $A_n \in \mathcal{K}_0$ such that $\|A_n - A\|)0$. Since the sequence of functions $\mathcal{A}_n(t,\mu)$ converges to $\mathcal{A}(t,\mu)$ uniformly on $\mathcal{M}$, the index $ind \; \mathcal{A}_n(t,\mu)$ stabilizes for sufficiently large n. We define $ind \; \mathcal{A}(t,\mu)$ by the formula

$$ind \; \mathcal{A}(t,\mu) = \lim \; ind \; \mathcal{A}_n(t,\mu) \; .$$

In this case we have

THEOREM 16.2. *Let* $A \in \mathcal{K}$ *and* $\mathcal{A}(t,\mu) \neq 0$ *on* $\mathcal{M}$. *Then* $A \in \Phi(\ell_2)$ *and*

$$Ind \; A = -ind \; \mathcal{A}(t,\mu) \; .$$

$\square$

We notice the following corollary of this theorem:

COROLLARY 16.2. *Let* $a, b \in PC(\Gamma_0)$. *Then there exists a function* $c \in PC(\Gamma_0)$ *such that* $T_a T_b - T_c \in \mathcal{T}(\ell_2)$ *if and only if* a *and* b *have no common points of discontinuity. If this condition is satisfied, then* $c = ab$.

PROOF. If $T_a T_b - T_c \in \mathcal{T}(\ell_2)$, then by equality (8.3)

$$
\begin{aligned}
&[a(t+0)\mu + a(t-0)(1-\mu)][b(t+0)\mu + b(t-0)(1-\mu)] \\
&\quad - c(t+0)\mu - c(t-0)(1-\mu) \equiv 0 \; .
\end{aligned}
\tag{16.4}
$$

Letting here $\mu = 0$ and then $\mu = 1$, we see that $a(t-0)b(t-0) = c(t-0)$ and $a(t+0)b(t+0) = c(t+0)$, i.e. $ab = c$. Now letting $c = ab$ and $\mu = \frac{1}{2}$ in (16.4), we get

$$[a(t+0) - a(t-0)][b(t+0) - b(t-0)] \equiv 0 \,,$$

which implies that at each point $t \in \Gamma_0$ at least one of functions a, b is continuous. The converse has been established in Lemma 8.2.

□

We shall also need

COROLLARY 16.3. $\Phi(\ell_2) \cap \mathcal{K}$ is dense in $\mathcal{K}$.

PROOF. Let $A \in \mathcal{K}$. Then there exists an operator B of the form $\sum \prod T_{a_{jk}}$ (with the norm $\|B - A\|$ sufficiently small), where a_{jk} are piecewise-smooth functions. Since the range of the function $\sum \prod a_{jk}(t, \mu)$ cannot fill a neighborhood of the point $\ell_0 = 0$, it follows that for every $\delta > 0$ one can find a $\lambda \in \mathbb{C}$ such that $|\lambda| < \delta$ and $B(t, \mu) \neq \lambda$ for all $(t, \mu) \in \Gamma \times [0, 1]$. The operator $C = B - \lambda I$ belongs to $\Phi(\ell_2) \cap \mathcal{K}$, and the norm $\|C - A\|$ is sufficiently small.

□

The results obtained above can be generalized to the matrix case.

THEOREM 16.3. *The algebra $\mathcal{K}$ of bounded linear operators $A \in \mathcal{L}(\ell_2^n)$ of the form $[A_{jk}]_{j,k=1}^n$, with $A_{jk} \in \mathcal{K}$, contains $T(\ell_2^n)$. The quotient algebra $\tilde{\mathcal{K}}(\ell_2^n)$ is isometrically isomorphic to the algebra $C^{n \times n}(\mathcal{M})$ of all continuous matrix-valued functions on the cylinder $\mathcal{M} = \Gamma_0 \times [0, 1]$. The operator $A \in \tilde{\mathcal{K}}$ is a Φ-operator in ℓ_2^n if and only if the matrix-valued function $\mathcal{A}(t, \mu) = [A_{jk}(t, \mu)]^n$ is nonsingular on $\mathcal{M}$. If this condition is satisfied, then*

$$Ind\ A = -ind\ det\ \mathcal{A}(t, \mu) \ .$$

The norm in the algebra $C^{n \times n}(\mathcal{M})$ is, by definition, $\|g\| = \max_{(t,\mu) \in \mathcal{M}} \|g(t, \mu)\|$, where for each point $(t, \mu) \in \mathcal{M}$, $\|g(t, \mu)\|$ designates the norm of the matrix $g(t, \mu)$ as an operator acting in $\mathbb{C}^n$. In other words,

$$\|g(t,\mu)\|^2 = \max_{\lambda \in \sigma(g(t,\mu)g^*(t,\mu))} \lambda \ .$$

PROOF. The fact that the algebras $\hat{\mathcal{K}}/\mathcal{T}(\ell_2^n)$ and $C^{n \times n}(\mathcal{M})$ are isomorphic follows from Theorem 16.1. Since they are both C^*-algebras and the norm of any element is equal to its spectral radius, the isomorphism between $\hat{\mathcal{K}}/\mathcal{T}(\ell_2^n)$ and $C^{n \times n}(\mathcal{M})$ is isometric. The assertion

$$(A \in \Phi(\ell_2^n)) \Longleftrightarrow (det[\mathcal{A}_{jk}(t,\mu)] \neq 0 \ \forall \ (t,\mu) \in \Gamma_0[0,1])$$

follows from Theorems 16.1 and 2.1. Finally, the index formula is a consequence of Theorems 3.1, 16.2 and Corollary 16.3.

$\square$

Thus far in this section we made essential use of the fact that $\mathcal{K}$ and $\hat{\mathcal{K}}$ are C^*-algebras. It is interesting to consider an algebra generated by Toeplitz operators which does not enjoy the C^*-property.

Let H_p $(1 < p < \infty)$ be the Hardy Banach space. As in Sec.10, we let h_p denote the Banach space, isometrically isomorphic to H_p, consisting of numerical sequences $\xi = \{\xi_n\}_{n=0}^\infty$ of Fourier coefficients of the functions from H_p. To each function $a \in PC(\Gamma_0)$ we assign the operator T_a in h_p defined by the Toeplitz matrix $[a_{j-k}]_{j,k=0}^\infty$, where a_k are the Fourier coefficients of a. Then T_a is the matrix representation of the operator $PaI \in \mathcal{L}(H_p)$ for $1 < p < \infty$, and hence $T_a \in \mathcal{L}(h_p)$. We let $\mathcal{K}_0$ denote the nonclosed subalgebra of $\mathcal{L}(h_p)$ generated by all operators T_a with $a \in PC(\Gamma_0)$. Let $\mathcal{K}_p$ $(1 < p < \infty)$ denote the closure of $\mathcal{K}_0$ in $\mathcal{L}(h_p)$.

LEMMA 16.4. *Let $a, b \in PC(\Gamma_0)$. The operator $K = T_a T_b - T_{ab}$ is compact in h_p if and only if the functions a and b have no common points of discontinuity.*

PROOF. Let u be the isometric mapping which sends each function $f \in H_p$ into the sequence $\{f_n\}_{n=0}^\infty \in h_p$ of its Fourier coefficients. The operator K can be represented as $K = uTu^{-1}$, where $T = PaPbP - PabP$; T is bounded in $L_p(\Gamma_0)$ for all $p \in (1, \infty)$. If

a and b have no common points of discontinuity, then by Lemma 16.2, $T \in \mathcal{T}(L_2(\Gamma_0))$. It follows from an interpolation theorem of Krasnosel'skii [36, Theorem 1] that T is compact in $L_p(\Gamma_0)$ for all $p \in (1, \infty)$. Conversely, if $T \in \mathcal{T}(L_{p_0}(\Gamma_0))$ for some $p_0 \in (1, \infty)$, then $T \in \mathcal{T}(L_2(\Gamma_0))$ (by the same interpolation theorem). Now the fact that a and b have no common points of discontinuity follows from Corollary 16.2.

□

In the same way we can use Lemma 16.3 to prove the following result:

LEMMA 16.5. *If* $a, b \in PC(\Gamma_0)$, *then* $T_a T_b - T_b T_a \in \mathcal{T}(h_p)$ *for all* $p \in (1, \infty)$.

□

It follows from this lemma that $\hat{\mathcal{K}}_p$ is a commutative Banach algebra, and hence $\hat{\mathcal{K}}_p \in \sigma_1$.

For $a \in PC(\Gamma_0)$ we let $a_p(t, \mu)$ denote the function

$$a_p(t, \mu) = a(t+0)f_p(t, \mu) + a(t-0)(1 - f_p(t, \mu)) ,$$

where $f_p(t, \mu)$ is defined by formula (9.1).

THEOREM 16.4.   *The compact space of maximal ideals of the algebra* $\hat{\mathcal{K}}_p$ *is homeomorphic to the cylinder* $\mathcal{M}$ *endowed with the topology defined by the neighborhoods (16.1). The value of a general multiplicative functional* $f_{t,\mu}$ *on the generators* $\hat{T}_a$ *is* $f_{t,\mu}(\hat{T}_a) = a_p(t, \mu)$.

To prove this theorem we need two lemmas.

LEMMA 16.6.   *Let*

$$A = \sum_{j=1}^{n} \sum_{k=1}^{m} T_{a_{jk}}$$

and

$$A_p(t, \mu) = \sum_{j=1}^{n} \sum_{k=1}^{m} (a_{jk})_p(t, \mu) .$$

Then

$$A \in \Phi(h_p) \iff A_p(t, \mu) \neq 0 \quad \forall \, (t, \mu) \in \Gamma_0 \times [0, 1] .$$

If this condition is satisfied, then

$$Ind \, A = -ind \, A_p(t, \mu) .$$

PROOF. One can proceed as in the case $p = 2$ considered above, passing to the linear dilation and using Theorem 4.4.

$\square$

LEMMA 16.7. *Let $A = \sum_{j=1}^{n} \prod_{k=1}^{m} T_{a_{jk}}$. Then*

$$\max_{(t,\mu)\in\Gamma\times[0,1]} |\mathcal{A}_p(t,\mu)| \leq \|\hat{A}\| . \tag{16.5}$$

PROOF. Suppose that for some p and A inequality (16.5) does not hold. Then there is a point $(t_0, \mu_0) \in \Gamma_0 \times [0,1]$ and an operator $K_0 \in \mathcal{T}(h_p)$ such that $\|A + K_0\|_p < |\mathcal{A}_p(t_0, \mu_0)|$. Set $B = \mathcal{A}_p(t_0, \mu_0)^{-1}A$ and $K = \mathcal{A}_p(t_0, \mu_0)^{-1}K_0$. Then $\|B + K\|_p < 1$ and hence $I - B \in \Phi(h_p)$. But this contradicts Lemma 16.6, since $(\mathcal{I} - \mathcal{B})_p(t_0, \mu_0) = 0$.

$\square$

From this lemma it follows, in particular, that the function $\mathcal{A}_p(t, \mu)$ does not depend on the way in which the operator $A \in \mathcal{K}_0$ is represented as a sum of products of Toeplitz operators. This in turn implies the multiplicativity and boundedness of every functional $f_{t,\mu}(\hat{A}) = \mathcal{A}_p(t, \mu)$ on the algebra $\hat{\mathcal{K}}_0$, and hence on the algebra $\hat{\mathcal{K}}_p$.

To complete the proof of Theorem 16.4 we must show that every multiplicative functional on $\hat{\mathcal{K}}_p$ is of the form $f_{t,\mu}$, with $(t, \mu) \in \Gamma_0 \times [0,1]$. Let f be an arbitrary multiplicative functional on $\hat{\mathcal{K}}_p$. We first show that there exists a point ζ_0, $|\zeta_0| = 1$, such that $f(\hat{T}_a) = a(\zeta_0)$ for all $a \in C(\Gamma_0)$. Suppose, on the contrary, that for every point $\tau \in \Gamma_0$ one can find a continuous function x_τ such that $f(\hat{T}_{x_\tau}) \neq x_\tau(\tau)$. We choose a neighborhood $U(\tau)$ of τ on Γ_0 such that $|x_\tau(t) - \alpha_\tau| \geq \delta_\tau > 0$ for all $t \in U(\tau)$, where $\alpha_\tau = f(\hat{T}_{x_\tau})$. Let $U(\tau_1), \ldots, U(\tau_n)$ be a finite cover of Γ_0. Then the function $y(\zeta) = \sum_{k=1}^{n}|x_{\tau_k}(\zeta)-\alpha_{\tau_k}|^2 \neq 0$, and hence $\hat{T}_y \hat{T}_{y^{-1}} = I$, i.e., $\hat{T}_y \in G\hat{\mathcal{K}}_p$, which contradicts the fact $\hat{T}_y \in Ker\, f$. We have thus established the existence of a point $\zeta_0 \in \Gamma_0$ with the desired property.

Now let us show that $f(\hat{T}_x) = x(\zeta_0)$ for every function $x \in PC(\Gamma_0)$ which is continuous at the point ζ_0. To this end we remark that every such x can be expressed as $x = y + az$, where $a \in PC(\Gamma_0)$, $y, z \in C(\Gamma_0)$, and $z(\zeta_0) = 0$. Indeed, it is readily checked that there is a continuous function y such that $y(\zeta_0) = x(\zeta_0)$, $(x - y)(t) \neq 0$ for $t \neq \zeta_0$, and $x - y$ is continuous in a neighborhood of ζ_0. We can now choose a function $z \in C(\Gamma_0)$ which coincides with $x - y$ in a neighborhood of ζ_0, and a function $a \in PC(\Gamma_0)$ such that $x - y = az$. For such a representation of x we have $f(x) = f(y) + f(a)f(z)$. By the preceding step of the proof, $f(z) = 0$ and $f(y) = y(\zeta_0) = x(\zeta_0)$, whence $f(x) = x(\zeta_0)$, as claimed.

We let $\lambda_p(\mu)$ (with $p \neq 2$) denote the circular arc in the complex plane described by the point $sin(\mu\theta)[sin\ \theta]^{-1}exp(i(\mu - 1)\theta)$, where $\theta = \pi(p - 2)/p$, as μ changes from 0 to 1; for $p = 2$ we put $\lambda_2(\mu) = [0, 1]$. Let $\chi \in PC(\Gamma_0)$ have the following properties: $\chi(\zeta_0 - 0) = 0$, $\chi(\zeta_0 + 0) = 1$; $\chi(\zeta)$ is continuous on $\Gamma_0 \backslash \{\zeta_0\}$; the range of χ coincides with the arc $\lambda_p(\mu)$. Then it follows from Theorem 10.4 that the spectrum of the element $\hat{T}_\chi$ in the algebra $\widehat{\mathcal{L}(h_p)}$ coincides with the arc $\lambda_p(\mu)$. Since the regular set of $\hat{T}_\chi$ is connected, the spectrum of $\hat{T}_\chi$ does not change on passing to the subalgebra $\hat{\mathcal{K}}_p$. This implies the existence of a point $\mu_0 \in [0, 1]$ such that $f(\hat{T}_\chi) = \lambda_p(\mu_0)$.

The final step of the proof is to show that $f(\hat{T}_x) = x_p(\zeta_0, \mu_0)$ for every function $x \in PC(\Gamma_0)$. Let $x \in PC(\Gamma_0)$ and let $y = x - c$, where the number c is chosen so that the function $a(t) = \chi(t)[y(\zeta_0 + 0) - y(\zeta_0 - 0)] + y(\zeta_0 - 0)$ is different from zero for all $t \in \Gamma_0$. It is readily verified that the function $b(t) = y(t)[(a(t)]^{-1}$ is continuous at the point ζ_0, and hence $f(\hat{T}_b) = b(\zeta_0)$. Then

$$f(\hat{T}_a) = \lambda_p(\mu_0)[y(\zeta_0 + 0) - y(\zeta_0 - 0)] + y(\zeta_0 - 0) = y_p(\zeta_0, \mu_0) ,$$

from which it follows that

$$f(\hat{T}_x) = f(\hat{T}_y) + c = f(\hat{T}_a) + c = y_p(\zeta_0, \mu_0) + c = x_p(\zeta_0, \mu_0) ,$$

as desired.

□

THEOREM 16.5. *Let $A \in K_p$. Then for A to be a Φ-operator in h_p it is necessary and sufficient that $A(t, \mu) \neq 0$ on $M = \Gamma_0 \times [0, 1]$. If this condition is satisfied, then*

$$Ind\ A = -ind\ A_p(t, \mu)\ .$$

PROOF. If $A_p(t, \mu) \neq 0$, then $\hat{A} \in G\hat{K}_p \Rightarrow \hat{A} \in G\widehat{\mathcal{L}(h_p)} \Rightarrow \hat{A} \in \Phi(h_p)$, which proves that the indicated condition is sufficient. To prove that it is also necessary we proceed by reduction ad absurdum. Suppose that $A \in \Phi(h_p) \cap K_p$, but $A_p(t_0, \mu_0) = 0$. Let $A_n \in K_0$ be a sequence such that $\|A_n - A\| \to 0$ as $n \to \infty$. Then by inequality (16.5), $(A_n)_p(t_0, \mu_0) \to 0$ as $n \to \infty$. Set $B_n = A_n - (A_n)_p(t_0, \mu_0)I$. Then $B_n \in K_0$, $B_n \in \Phi(h_p)$ for sufficiently large n, and $(B_n)_p(t_0, \mu_0) = 0$, which contradicts Lemma 16.6. The index formula is established in the same way as in the case $p = 2$.

□

THEOREM 16.6. *Let $A = [A_{jk}]_1^n \in K_p^{n \times n}$. Then for A to be a Φ-operator in h_p^n it is necessary and sufficient that $det[(A_{jk})_p(t, \mu)] \neq 0$ on $\Gamma_0 \times [0, 1]$. If this condition is satisfied, then*

$$Ind\ A = -ind\ det[(A_{jk})_p(t, \mu)]\ .$$

The proof is analogous to that of Theorem 16.3.

□

CHAPTER V

BANACH ALGEBRAS GENERATED BY
SINGULAR INTEGRAL OPERATORS WITH
PIECEWISE-CONTINUOUS COEFFICIENTS

In this chapter we show that the algebras generated by singular integral operators with piecewise-continuous coefficients cannot be endowed with a scalar symbol. On such algebras we introduce matrix symbols and study their properties.

17. MATRIX SYMBOLS ON ALGEBRAS OF
SINGULAR INTEGRAL OPERATORS

If the algebra $K \subset \mathcal{L}(E))$ belongs to the class σ_1, then on $K^{n \times n} \subset \mathcal{L}(E)^n)$ one can introduce a matrix symbol (see, e.g., Theorems 16.3 and 16.6). A matrix symbol arises also when one extends K by adjoining shift operators (see, e.g., Sec.19). It seems, however, less natural that a matrix symbol should exist on an algebra generated by singular integral operators with piecewise-continuous coefficients. It is to this question that we devote the present section.

Let Γ be a simple closed Lyapunov contour and $E = L_p(\Gamma)$, $1 < p < \infty$. We let K_p denote the smallest Banach subalgebra of $\mathcal{L}(E))$ that contains all the operators $(H\varphi)(t) = h(t)\varphi(t)$ of multiplication by functions $h \in PC(\Gamma)$, the operator S of singular integration along Γ, and all the operators $T \in \mathcal{T}(E)$. Can we introduce a symbol on K_p? According to Theorem 14.3, if $K_p \in \sigma_1$, then $T = AB - BA$ is a Φ-admissible perturbation for the operators in K_p. In particular, $A = \lambda T + aS - SaI \in \Phi(E)$ for all $\lambda \neq 0$ and all

$a \in PC(\Gamma)$. Equivalently,

$$\tilde{A} = \begin{bmatrix} I & aI \\ S & \lambda I + aS \end{bmatrix} = \begin{bmatrix} 1 & a \\ 1 & \lambda + a \end{bmatrix} P + \begin{bmatrix} 1 & a \\ -1 & \lambda - a \end{bmatrix} Q \tag{17.1}$$

is a Φ-operator in E^2 (Theorem 2.3). Also, $\tilde{A}$ is a singular integral operator with matrix coefficients. By Theorem 10.1, the matrix-valued function

$$c = \begin{bmatrix} 1 & a \\ -1 & \lambda - a \end{bmatrix}^{-1} \begin{bmatrix} 1 & a \\ 1 & \lambda + a \end{bmatrix}$$

must be p-nonsingular. Since

$$c = \lambda^{-1} \begin{bmatrix} \lambda - 2a & -2a^2 \\ 2 & 2a + \lambda \end{bmatrix} ,$$

$$det\ c_p(t, \mu) = \lambda^{-2} f(t, \mu)(1 - f(t, \mu))[a(t + 0) - a(t - 0)]^2 + 1 .$$

If the function a has a point of discontinuity t_0, then for the choice $\lambda = [f(t_0, \mu_0)(1 - f(t_0, \mu_0))]^{1/2}[a(t_0 + 0) - a(t_0 - 0)]$ we obtain $det\ c_p(t_0, \mu_0) = 0$. Therefore, if the function $a \in PC(\Gamma)$ has at least one (nonremovable) point of discontinuity, then $aS - SaI$ is not a Φ-admissible perturbation for the algebra K_p, i.e., $K_p \notin \sigma_1$.

We let $\mathcal{V}$ denote the linear manifold consisting of the operators $A = aP + bQ$ with $a, b \in PC(\Gamma)$. The following criterion for an operator $A \in \mathcal{V}$ to be a Fredholm operator has been obtained in [50(9)] (see Sec.9):

$$\begin{aligned} aP + bQ \in \Phi(L_p(\Gamma)) &\Longleftrightarrow \\ \left(\begin{aligned} a(t + 0)b(t - 0)f(t, \mu) + a(t - 0)b(t + 0)(1 - f(t, \mu)) &\neq 0 \\ \forall\ (t, \mu) \in \Gamma \times [0, 1] \end{aligned} \right) & . \end{aligned} \tag{17.2}$$

We write this condition in the form $\varphi(t, \mu) \neq 0$. Then the mapping $aP + bQ \to \varphi(t_0, \mu_0)$ is linear on $\mathcal{V}$. The expression obtained for φ suggests that this function may be expressed as

$$\varphi(t, \mu) = det[\alpha_{jk}(\mu)a(t + 0) + \beta_{jk}(\mu)a(t - 0)]_{j,k=1}^2 \tag{17.3}$$

with appropriate $\alpha_{jk}(\mu)$ and $\beta_{jk}(\mu)$. The matrix on the right-hand side of (17.3) can be obtained by representing the operator $A = aP + bQ$ in the matrix form corresponding to the decomposition $E = PE + QE$:

$$A = \begin{bmatrix} PaP & PbQ \\ QaP & QbQ \end{bmatrix} .$$

According to this rule, the matrix

$$diag(\alpha_{11}(\mu)a(t+0) + \beta_{11}(\mu)a(t-0), 1)$$

should correspond to the operator $PaP + Q$. Recalling the criterion for $PaP + Q$ to be a Fredholm operator, we are naturally led to the choice

$$\alpha_{11}(\mu)a(t+0) + \beta_{11}(\mu)a(t-0) = a_p(t,\mu) .$$

Analogous use of the condition for $P + QaQ$ to be a Fredholm operator in conjunction with the requirement that to the adjoint operator there should correspond the conjugate transpose matrix leads finally to the mapping

$$aP + bQ \xrightarrow{\gamma}$$

$$\begin{bmatrix} a(t+0)f(t,\mu) + a(t-0)(1 - f(t,\mu)) & \sqrt{f(t,\mu)(1 - f(t,\mu))}(b(t+0) - b(t-0)) \\ \sqrt{f(t,\mu)(1 - f(t,\mu))}(a(t+0) - a(t-0)) & b(t+0)(1 - f(t,\mu)) + b(t-0)f(t,\mu) \end{bmatrix} .$$

The mapping γ $(= \gamma_{t,\mu})$ is linear on $\mathcal{V}$ and provides a test for the Fredholmness of the operator $aP + bQ$. One is naturally led to asking whether γ can be extended from the set $\mathcal{V}$ of generators of the algebra K_p to a homomorphism $\gamma_{t,\mu} : K_p \to \mathbb{C}^{2\times2}$ that serves as a matrix symbol. We give an affirmative answer below.

Another approach to the construction of a matrix symbol on an algebra generated by singular integral operators with piecewise-continuous coefficients is discussed in the following example.

Let $E = L_p(\mathbb{R})$, $CP_0(\mathbb{R})$ the set of functions which are continuous at every point $t \neq 0$ and have finite limits for $t \to \pm 0$ and $t \to \pm\infty$, $S_{\mathbb{R}}$ the operator of singular integration along $\mathbb{R}$, and K_p the Banach algebra generated by $S_{\mathbb{R}}$ and all the operators of multiplications by functions $a \in CP_0(\mathbb{R})$. We let K_p^+ denote the subalgebra of $\mathcal{L}(L_p(\mathbb{R}_+))$ generated by the singular integral operators whose coefficients are continuous on $[0, \infty]$.

Since the algebra $\hat{K}_p^+$ is commutative, $K_p^+ \in \sigma_1$. Let $v : L_p(\mathbb{R}) \to L_p^2(\mathbb{R}_+)$ denote the isometric mapping defined by the formula $(v\varphi)(t) = (\varphi(t), \varphi(-t))$. Then

$$vHv^{-1} = \begin{bmatrix} H_1 & 0 \\ 0 & H_2 \end{bmatrix} \quad \text{and} \quad vS_{\mathbb{R}}v^{-1} = \begin{bmatrix} S & -R \\ R & -S \end{bmatrix},$$

where

$$
\begin{aligned}
(H\varphi)(t) &= h(t)\varphi(t) &, \; t \in \mathbb{R} &, \\
(H_1\varphi)(t) &= h(t)\varphi(t) &, \; t > 0 &, \\
(H_2\varphi)(t) &= h(-t)\varphi(t) &, \; t > 0 &,
\end{aligned}
$$

and

$$
\left.
\begin{aligned}
(S\varphi)(t) &= \frac{1}{\pi i} \int_0^\infty \frac{\varphi(\tau)d\tau}{\tau - t}, \\
(R\varphi)(t) &= \frac{1}{\pi i} \int_0^\infty \frac{\varphi(\tau)d\tau}{\tau + t}.
\end{aligned}
\right\}
\tag{17.4}
$$

In Example 13.6 it was shown that the operator R belongs to K_p^+ (actually, it belongs even to the algebra generated by the singular integral operators with constant coefficients). Therefore,

$$vK_pv^{-1} \subset (K_p^+)^{2 \times 2}.$$

As we remarked above, $\hat{K}_p^+$ is a commutative algebra and $K_p^+ \in \sigma_1$. Let $\{\gamma_M\}_{M \in \alpha}$ be a family of homomorphisms which define a symbol on K_p^+. For each operator $A \in K_p$ we set $\tilde{\gamma}_M(A) = [\gamma_M(A_{jk})]_{j,k=1}^2$ where $vAv^{-1} = [A_{jk}]_{j,k=1}^2$.

THEOREM 17.1. Let $A \in K_p$. Then A is a Φ-operator in $E = L_p(\mathbb{R})$ if and only if $\det \tilde{\gamma}_M(A) \neq 0$ for every $\alpha \in M$.

PROOF. By Theorem 2.1,

$$[A_{jk}]_{j,k=1}^2 \in \Phi(L_p^2(\mathbb{R}_+)) \iff \det[A_{jk}] \in \Phi(L_p(\mathbb{R}_+))$$

$$\iff \det[\gamma_M(A_{jk})] \neq 0 \quad \forall M \in \alpha.$$

But $\gamma_M(det[A_{jk}]) = det[\gamma_M(A_{jk})]$, and hence $A \in \Phi(E) \Longleftrightarrow det \, \tilde{\gamma}_M(A) \neq 0 \quad \forall \, M \in \alpha$.

$\square$

Suppose now that the contour Γ consists of n rays: $\Gamma = \bigcup_{k=1}^{n} \Gamma_k$, where $\Gamma_m = \{\varepsilon_k x : x \in \mathbb{R}_+, \varepsilon_k \in \mathbb{C}, |\varepsilon_k| = 1\}$. We let $PC_0(\Gamma)$ denote the set of functions which are continuous on $\Gamma \backslash \{0\}$ and have finite limits as $t \to 0$ and $t \to \infty$ along each ray Γ_m. Let $K_p \subset \mathcal{L}(L_p(\Gamma))$ be the algebra generated by the singular operators with coefficients from $PC_0(\Gamma)$. We shall take $\varepsilon_1 = 1$, so that $\Gamma_1 = \mathbb{R}_+$. Let $u : L_p(\Gamma) \to L_p^n(\Gamma_1)$ denote the isometric mapping defined by the rule $u\varphi = (\varphi_1, \ldots, \varphi_n)$, where $\varphi_k(t) = \varphi(\varepsilon_k t)$ for $t \geq 0$ and $k = 1, \ldots, n$. Then

$$uHu^{-1} = \begin{bmatrix} H_1 & & 0 \\ & \ddots & \\ 0 & & H_n \end{bmatrix}$$

and

$$uS_\Gamma u^{-1} = [R_{jk}]_{j,k=1}^n \, ,$$

where

$$(H\varphi)(t) = h(t)\varphi(t) \, , \qquad t \in \Gamma \, ,$$

$$(H_k\varphi)(t) = h(\varepsilon_k t)\varphi(t) \, , \qquad t \in \Gamma_1 \, ,$$

and

$$(R_{jk}\varphi)(t) = \frac{1}{\pi i} \int_0^\infty \frac{\varphi(s)\,ds}{\varepsilon - \varepsilon_k^{-1}\varepsilon_j t} \, . \tag{17.5}$$

In Example 13.6 it was shown that $R_{jk} \in K_p^+$. Therefore, $uK_p u^{-1} \subset (K_p^+)^{n \times n}$. As in Theorem 17.1, one can show for operators $A \in K_p$ that

$$A \in \Phi(E) \Longleftrightarrow det[\gamma_M(A_{jk})] \neq 0 \quad \forall \, M \in \alpha \, , \tag{17.6}$$

where $uAu^{-1} = [A_{jk}]_{j,k=1}^n$. Therefore, the family $\tilde{\gamma}_M(A) = [\gamma_M(A_{jk})]_{j,k=1}^n$, $M \in \alpha$, defines a matrix symbol on K_p.

18. EXTENSION OF PRESYMBOLS

Let $\mathcal{V}$ be a linear manifold in $\mathcal{L}(E)$ which contains the identity operator I. We let $\mathcal{K}_0$ and $\mathcal{K}$ denote the set of operators of the form

$$A = \sum_{j=1}^{\ell} A_{j1} A_{j2} \cdot \ldots \cdot A_{jr} \qquad (\ell \in \mathbb{N}, \ A_{jk} \in \mathcal{V}) \tag{18.1}$$

and the closure of $\mathcal{K}_0$ in the algebra $\mathcal{L}(E)$, respectively.

Definition 18.1. We call $\mathcal{K}$ an *algebra with an n-symbol* if there exists a family $\{\gamma_M\}_{M \in \alpha}$ of homomorphisms $\gamma_M : \mathcal{K} \to \mathbb{C}^{\ell \times \ell}$, where $\ell = \ell(M) \le n$, such that if the operator $A \in \mathcal{K}$, then

$$A \in \Phi(E) \iff (det \ \gamma_M(A) \ne 0 \quad \forall \ M \in \alpha). \tag{18.2}$$

Definition 18.2. We say that an *n-presymbol* is defined on the linear manifold $\mathcal{V}$ if there is given a family $\{\gamma_M\}_{M \in \alpha}$ of linear maps $\gamma_M : \mathcal{V} \to \mathbb{C}^{\ell \times \ell}$, where $\ell = \ell(M) \le n$, such that $\gamma_M(I) = e$ (the identity matrix) and for every $s \in \mathbb{N}$ and operators $A_{jk} \in \mathcal{V}$, $j, k = 1, \ldots, s$,

$$[A_{jk}]_{j,k=1}^n \in \Phi(E^s) \iff (det \ [\gamma_M(A_{jk})]_{j,s=1}^s \ne 0 \quad \forall \ M \in \alpha). \tag{18.3}$$

To each operator A of the form (18.1) we assign the family of matrices

$$\tilde{\gamma}_M(A) = \sum_{j=1}^{\ell} \gamma_M(A_{j1}) \gamma_M(A_{j2}) \ldots \gamma_M(A_{jr}), \qquad M \in \alpha. \tag{18.4}$$

We shall be concerned with the following questions:

1) Does $\tilde{\gamma}_M(A)$ depend on the representation of the operator A in the form (18.1)?

2) Does assertion (18.2) remain valid for the operator (18.1) if one replaces γ_M by $\tilde{\gamma}_M$?

3) Is it possible to extend $\tilde{\gamma}_M$ to a homomorphism of the algebra $\mathcal{K}$ into $\mathbb{C}^{\ell \times \ell}$?

4) If the answer to 3) is positive, does the family $\{\tilde{\gamma}_M\}_{M \in \alpha}$ define a symbol on $\mathcal{K}$?

We first answer the second question. Even if $\tilde{\gamma}_M$ is not correctly defined on $\mathcal{K}_0$, the family $\{\tilde{\gamma}_M\}_{M \in \alpha}$ provides a criterion for the Fredholmness of the operators $A \in \mathcal{K}_0$, as follows from the next result.

THEOREM 18.1. *Suppose the operator $A \in \mathcal{K}_0$ is represented somehow in the form* (18.1), *and let $\tilde{\gamma}_M(A)$ be the matrices defined by the rule* (18.4). *Then*

$$A \in \Phi(E) \Longleftrightarrow (det \ \tilde{\gamma}_M(A) \neq 0 \quad \forall \ M \in \alpha). \qquad (18.5)$$

PROOF. Let $\Xi(A) = [B_{jk}]_{j,k=1}^s$, $B_{jk} \in \mathcal{V}$, be the linear dilation of A corresponding to representation (18.1). Since $\gamma_M(\lambda I) = \lambda e$, it follows that $\Xi(\tilde{\gamma}_M(A)) = [\gamma_M(B_{jk})]_{j,k=1}^s$. By Theorem 2.4, $A \in \Phi(E) \Longleftrightarrow \Xi(A) \in \Phi(E^s)$. By (18.3), $\Xi(A) \in \Phi(E^s) \Longleftrightarrow (det \ \tilde{\gamma}_M(A) \neq 0 \quad \forall \ M \in \alpha)$. Now (18.5) follows from the equality $det \ \Xi(\tilde{\gamma}_M(A)) = det \ \tilde{\gamma}_M(A)$.

$\square$

The answer to the first question is, generally speaking, negative.

For example, consider in $L_2^2(0,1)$ the set $\mathcal{V}$ of operators of multiplication by diagonal matrix-functions $A = \mathrm{diag}(a_1, a_2)$, where $a_1', a_2' \in C(0,1)$. Let

$$\gamma_s(A) = \begin{bmatrix} a_1(s) & a_1'(s) \\ 0 & a_2(s) \end{bmatrix}, \qquad s \in [0,1]. \qquad (18.6)$$

It is readily verified that $\{\gamma_s\}_{s \in [0,1]}$ is a 2-presymbol on $\mathcal{V}$. However,

$$\begin{bmatrix} t & 0 \\ 0 & 0 \end{bmatrix} \begin{bmatrix} 0 & 0 \\ 0 & 1 \end{bmatrix} = \begin{bmatrix} 0 & 0 \\ 0 & 0 \end{bmatrix},$$

whereas

$$\gamma_s \left(\begin{bmatrix} t & 0 \\ 0 & 0 \end{bmatrix} \right) \gamma_s \left(\begin{bmatrix} 0 & 0 \\ 0 & 1 \end{bmatrix} \right) = \begin{bmatrix} s & 1 \\ 0 & 0 \end{bmatrix} \cdot \begin{bmatrix} 0 & 0 \\ 0 & 1 \end{bmatrix} = \begin{bmatrix} 0 & 1 \\ 0 & 0 \end{bmatrix} \neq 0.$$

Therefore, $\tilde{\gamma}_s(A)$ depends on representation (18.1) of the operator A. This difficulty can be however circumvented as follows.

Let $\sum$ and $\sum_M$ denote the set of all representations (18.1) and the corresponding set of right-hand sides of formula (18.4), respectively. Then obviously $\sum_M$ is a subalgebra of $\mathbb{C}^{\ell(M)\times\ell(M)}$. By Wedderburn's theorem [68], $\sum_M$ decomposes into the direct sum $\sum_M^0 + \mathcal{R}_M$ of its radical $\mathcal{R}_M$ and a semisimple subalgebra $\sum_M^0$. Let $\nu_M : \sum_M \to \sum_M^0$ be the natural homomorphism and set $\tau_M = \nu_M \tilde{\gamma}_M$. Then

$$(\det \tilde{\gamma}_M(A) \neq 0) \Longleftrightarrow (\det \tau_M(A) \neq 0)$$

and hence

$$A \in \Phi(E) \Longleftrightarrow (\det \tau_M(A) \neq 0 \quad \forall \ M \in \alpha).$$

Let us show that $\tau_M(A)$ does not depend on the particular representation (18.1) of A. In fact, let $\mathcal{S}_M$ denote the set of matrices

$$\sum_{j=1}^{\ell} \tau_M(A_{j1})\tau_M(A_{j2})\ldots\tau_M(A_{jr})$$

corresponding to all possible representations

$$0 = \sum_{j=1}^{\ell} A_{j1}A_{j2}\ldots A_{jr}$$

of the zero operator. We show that $\mathcal{S}_M = \{0\}$. To this end it suffices to verify that $\mathcal{S}_M$ is the radical of $\sum_M^0$. Let $c \in \mathcal{S}_M$ and $b \in \sum_M^0$. Then there exist representations

$$0 = \sum \prod C_{jk} \quad \text{and} \quad B = \sum \prod B_{jk}$$

(the order of multiplication of C_{jk} and B_{jk} is important!) such that

$$\sum \prod \tau_M(C_{jk}) = c \quad \text{and} \quad \sum \prod \tau_M(B_{jk}) = b.$$

This implies that there exists a representation (18.1) of the identity operator, for which $e + cb = \tau_M(I)$, and hence $\det (e + cb) \neq 0$. Every subalgebra of $\mathbb{C}^{\ell\times\ell}$, and hence $\sum_M^0$, too, is inverse-closed. Therefore, $e + cb \in \sum_M^0$, which shows that c belongs to the radical of the algebra $\sum_M^0$. We have thus proved

THEOREM 18.2. *Suppose that an n-presymbol $\{\gamma_M\}_{M \in \alpha}$ is defined on $\mathcal{V}$. If the algebra $\sum_M$ is semisimple, then for each operator $A \in \mathcal{K}_0$ the matrix $\tilde{\gamma}_M(A)$ does not depend on the representation* (18.1) *of A.*

□

This result admits the following corollaries, in which it is assumed that $\sum_M$ is semisimple:

COROLLARY 18.1. *$\tilde{\gamma}_M$ is a homomorphism of algebra $\mathcal{K}_0$.*

□

COROLLARY 18.2. *If $T \in \mathcal{K}_0 \cap \mathcal{T}(E)$, then $\tilde{\gamma}_M(T) = 0$.*

PROOF. If the operator T is compact, then $I - AT \in \Phi(E)$ for all $A \in \mathcal{K}_0$, and hence $det\ (e - \tilde{\gamma}_M(A)\tilde{\gamma}_M(T)) \neq 0$. Thus, $\tilde{\gamma}_M(T)$ belongs to the radical of the semisimple algebra $\tilde{\gamma}_M(\mathcal{K}_0)$, i.e., $\tilde{\gamma}_M(T) = 0$.

□

We shall now examine a number of examples.

Example 18.1. Let Γ be a closed Lyapunov contour, $E = L_p(\Gamma, \rho)$ (with $1 < p < \infty$, $\rho(t) = \prod |t - t_k|^{\beta_k}$, $t_k \in \Gamma$, $-1 < \beta_k < p - 1$), and let $\mathcal{V}$ be the set of all operators of the form $aP + bQ$, where a, b are piecewise-continuous functions, $P = (I + S)/2$, $Q = (I - S)/2$, and S is the operator of singular integration along Γ. It follows from Theorem 4.3 that the family of maps $\gamma_{t,\mu}(aP + bQ) = \mathcal{A}_{p,\rho}(t, \mu)$, $(t, \mu) \in \Gamma \times [0, 1]$, defined by formula (10.8), defines a 2-presymbol on $\mathcal{V}$. For every choice of $\mu \neq 0, 1$ and $t \in \Gamma$, $\gamma_{t,\mu}(\mathcal{V}) = \mathbb{C}^{2 \times 2}$, whereas $\gamma_{1,\mu}(\mathcal{V})$ and $\gamma_{0,\mu}(\mathcal{V})$ each coincide with the set of all diagonal 2×2 matrices. This implies that $\tilde{\gamma}_{t,\mu}(\mathcal{V})$ is a semisimple algebras for every point $M = (t, \mu) \in \Gamma \times [0, 1]$.

According to Theorem 18.2, $\gamma_{t,\mu}$ admits an extension to a homomorphism of the algebra $\mathcal{K}_0$ into $\mathbb{C}^{2 \times 2}$, and the family of homomorphisms $\{\gamma_{t,\mu}\}_{(t,\mu) \in \Gamma \times [0,1]}$ defines a 2-symbol on $\mathcal{K}_0$ (concerning the extension of the symbol to $\mathcal{K}$ see below).

Example 18.2. Suppose that Γ is made of a finite number of simple closed and open Lyapunov contours with no common points. Let $E = L_p(\Gamma, \rho)$ and let $\mathcal{V}$ denote the set of all operators $aI + bS$ with $a, b \in C(\Gamma)$. Set

$$\gamma_{t,z}(aI + bS) = a(t) + b(t)\Omega_{p,\rho}(t, z)$$

where $\Omega_{p,\rho}(t, z)$ is the function defined by formula (10.11), and the point (t, z) runs through the spatial curve $\Lambda(\Gamma)$ described in Sec.10. It follows from Theorem 10.5 that the family $\{\gamma_{t,z}\}_{(t,z)\in\Lambda(\Gamma)}$ defines a 1-presymbol on $\mathcal{V}$. Since the algebra $\gamma_{t,z}(\mathcal{V})$ is semisimple, each $\gamma_{t,z}$ extends to a multiplicative functional on $\mathcal{K}_0$, and the family $\{\gamma_{t,z}\}_{(t,z)\in\Lambda(\Gamma)}$ defines a 1-symbol on $\mathcal{K}_0$.

Example 18.3. Let $E = L_2(\Gamma_0)$, and let $\mathcal{V}$ denote the set of operator of multiplication by functions $a \in PC(\Gamma_0)$. Since $\mathcal{V}$ is a subset of the linear manifold of Example 18.1 (for $\Gamma = \Gamma_0$ and $\rho(t) \equiv 1$), we can define a 2-presymbol on $\mathcal{V}$ by the same rule:

$$\gamma_{t,z}(aI) = \begin{bmatrix} a(t+0)\mu + a(t-0)(1-\mu) & \sqrt{\mu(1-\mu)}(a(t+0) - a(t-0)) \\ \sqrt{\mu(1-\mu)}(a(t+0) - a(t-0)) & a(t+0)(1-\mu) + a(t-0)\mu \end{bmatrix} . \quad (18.7)$$

In Example 18.1 we showed that $\gamma_{t,\mu}$ extends to a homomorphism on $\mathcal{K}_0$. Since $\mathcal{V}$ is a commutative subalgebra of $\mathcal{K}_0$, the set of matrices $\{\gamma_{t,\mu}(aI): a \in \mathcal{V}\}$ is a commutative algebra for every choice of $(t, \mu) \in \Gamma_0 \times [0, 1]$. If the image of $\gamma_{t,\mu}$ is semisimple, then it must be isomorphic to an algebra of diagonal matrices (see [68]). Let us show that this is indeed the case. In fact, $\gamma_{t,0}(aI) = \operatorname{diag}(a(t - 0), a(t + 0))$, $\gamma_{t,1}(aI) = \operatorname{diag}(a(t + 0), a(t - 0))$, and for $\mu \neq 0, 1$

$$\begin{bmatrix} \mu & \sqrt{\mu(1-\mu)} \\ \mu - 1 & \sqrt{\mu(1-\mu)} \end{bmatrix} \gamma_{t,\mu}(aI) \begin{bmatrix} 1 & -1 \\ \sqrt{(1-\mu)/\mu} & \sqrt{\mu/(1-\mu)} \end{bmatrix} = \begin{bmatrix} a(t+0) & 0 \\ 0 & a(t-0) \end{bmatrix} .$$

$$(18.8)$$

Remark 18.1. In each of the examples considered above one can arrange that the image of γ_M coincides with $\mathbb{C}^{\ell \times \ell}$ (where $\ell = \ell(M)$) for every $M \in \alpha$. In Example 18.3 one can take $\gamma_{t,1}(aI) = a(t + 0)$ and $\gamma_{t,-1}(aI) = a(t - 0)$; then M runs through the set $\Gamma \times \{-1, 1\}$. In Example 18.1, $\operatorname{Im} \gamma_{t,\mu} = \mathbb{C}^{2 \times 2}$ for all $\mu \neq 0, 1$, and the homomorphisms $\gamma_{t,1}$ and $\gamma_{t,0}$ may be replaced by the four homomorphisms defined on operators $A = aP + bQ$

by the formulas

$$\gamma_1(A) = a(t+0), \qquad \gamma_2(A) = a(t-0),$$

$$\gamma_3(A) = b(t+0), \qquad \gamma_4(A) = b(t-0).$$

Then $Im\ \gamma_j = \mathbb{C}$, $j = 1, 2, 3, 4$.

One can proceed similarly in the general case. If the algebra $Im\ \gamma_M$ is semisimple (which we already achieved), then it is a direct sum of full matrix algebras: $Im\ \gamma_M = \mathbb{C}^{k_1 \times k_1} \dotplus \ldots \dotplus \mathbb{C}^{k_r \times k_r}$. Let π_j denote the natural homomorphism $Im\ \gamma_M \to \mathbb{C}^{k_j \times k_j}$. We set $\gamma_{M_j} = \pi_j \gamma_M$ and then replace every homomorphism γ_M by the corresponding collection $\{\gamma_{M_j}\}$, $j = 1, \ldots, r(M)$. We thus obtain an m-symbol ($m \leq n$) on $\mathcal{K}_0$ (which we continue to denote by $\{\gamma_M\}$) such that $Im\ \gamma_M = \mathbb{C}^{\ell \times \ell}$, where ($\ell = \ell(M)$).

THEOREM 18.3. *Suppose that $\{\gamma_M\}_{M \in \alpha}$ defines an n-symbol on $\mathcal{K}$ and $Im\ \gamma_M = \mathbb{C}^{\ell(M) \times \ell(M)}$. Let $\gamma_M(A) = [a_{jk}(M)]$. Then there exist constants $\beta(M)$ such that $|a_{jk}(M)| \leq \beta(M)\|\hat{A}\|$ for all $A \in \mathcal{K}_0$ ($\beta(M)$ are independent of A).*

PROOF. Fix the homomorphism γ_M. Since $Im\ \gamma_M = \mathbb{C}^{\ell(M) \times \ell(M)}$, there exist operators $R_{jk} \in \mathcal{K}_0$, $j, k = 1, \ldots, \ell$, such that all the entries r_{pq} of the matrix $\gamma_M(R_{jk})$ are zero except for $r_{jk} = 1$ (i.e., $\gamma_M(R_{jk})$ are the matrix units). Let $A \in \mathcal{K}_0$. Then

$$\gamma_M(R_{1j}AR_{k1} - \lambda I) = diag(a_{jk} - \lambda, -\lambda, \ldots, -\lambda).$$

Hence, for $\lambda = a_{jk}(M)$, the operator $R_{1j}AR_{k1} - \lambda I$ is not Fredholm. Consequently, $a_{jk}(M)$ belongs to the spectrum of the element $\hat{R}_{1j}\hat{A}R_{k1}$ in the algebra $\widehat{\mathcal{L}(E)}$, and hence $|a_{jk}(M)| \leq \|\hat{R}_{1j}\| \cdot \|\hat{R}_{k1}\|\|\hat{A}\|$.

$\square$

COROLLARY 18.3. *Suppose that the family $\{\gamma_M\}_{M \in \alpha}$ defines a 1-symbol on $\mathcal{K}_0$. Then $|\gamma_M(A)| \leq \|\hat{A}\|$ for every $M \in \alpha$.*

$\square$

In Examples 18.2 and 18.3 $n = 1$, and so $|\gamma_M(A)| \leq \|\hat{A}\|$. In Example 18.1 we also have $|\gamma_{t,1}(A)| \leq \|\hat{A}\|$ and $|\gamma_{t,0}(A)| \leq \|\hat{A}\|$. Now let $\mu_0 \neq 0, 1$ and $t_0 \in \Gamma$. Pick an arbitrary function $a \in PC(\Gamma_0)$ (which may depend on μ_0) such that $(a(t_0 + 0) - a(t_0 - 0))\{f(t_0, \mu_0)(1 - f(t_0, \mu_0))\}^{1/2} = 1$. Then we may take $R_{12} = PaQ$,

$R_{21} = QaP$, $R_{11} = P$, and $R_{22} = Q$. We remark that, for example, for $p = 2$, $\rho(t) \equiv 1$, and $\mu_0 = 1/2$ we obtain in this way the constraint $a(t_0 + 0) - a(t_0 - 0) = 2$, which can be satisfied by choosing a function a which takes only the two values ± 1. Since under these circumstances $\|\hat{P}\| = \|\hat{Q}\| = \|aI\| = 1$, it follows that $|a_{jk}(t_0, 1/2)| \leq \|\hat{A}\|$.

Under the assumptions of Theorem 18.3, every homomorphism which gives an n-symbol on $\mathcal{K}_0$ can be extended by continuity to the algebra $\mathcal{K} = \bar{\mathcal{K}}_0$. One is naturally led to asking if the family $\{\gamma_M\}_{M \in \alpha}$ defines a symbol on $\mathcal{K}$.

LEMMA 18.1. *Suppose that the family* $\{\gamma_M\}_{M \in \alpha}$ *defines an n-symbol on* $\mathcal{K}_0$ *and Im* $\gamma_M = \mathbb{C}^{\ell \times \ell}$ *(with* $\ell = \ell(M)$*). Then* $\det \gamma_M(A) \neq 0$ *for every operator* $A \in \Phi(E) \cap \mathcal{K}$, *where* $\mathcal{K} = \bar{\mathcal{K}}_0$, *the closure of* $\mathcal{K}_0$.

PROOF. Let $A \in \Phi(E) \cap \mathcal{K}$ and suppose that $\det \gamma_M(A) = 0$ for some $M \in \alpha$. There exists a sequence of operators $A_n \in \mathcal{K}_0$ such that $\|A_n - A\| \to 0$. By Theorem 18.3, $\det \gamma_M(A_n) \to 0$ and hence there is a sequence of eigenvalues λ_n of the matrices $\gamma_M(A_n)$ such that $\lambda_n \to 0$. Set $B_n = A_n - \lambda_n I$. Since $\|A - B_n\| \to 0$, it follows that $B_n \in \Phi(E)$ for sufficiently large n, whereas $\det \gamma_M(B_n) = 0$.

$\square$

We shall now give an example which shows that the extension of a presymbol $\{\gamma_M\}_{M \in \alpha}$ from $\mathcal{K}_0$ to $\mathcal{K}$ is not necessarily a symbol.

Example 18.4. Let $\mathcal{V}$ denote the set of operators of multiplication: $(A\varphi)(t) = a(t)\varphi(t)$, $\varphi \in L_2(0,1)$, where a is an arbitrary piecewise – constant function on $[0, 1]$. Since every such function a has only a finite number of points of discontinuity, the condition $a(t) \neq 0$ for every $t \in [0,1]$ is necessary and sufficient for A to belong to $\Phi(E)$. In this example $\mathcal{K}_0 = \mathcal{V}$ and a 1-symbol on $\mathcal{K}_0$ is defined by the family of homomorphisms $\gamma_t(A) = a(t)$. Let us show that this family does not define a 1-symbol on $\mathcal{K} = \bar{\mathcal{K}}_0$. Set $a_0(t) = 2^{-n}$ for $2^{-n-1} < t \leq 2^{-n}$, $n = 0, 1, \ldots$, $a_0(0) = 1$, and $A_0 = a_0 I$. It is readily verified that $A_0 \in \mathcal{K}$ and $\gamma_t(A_0) = a_0(t) \neq 0$ for all $t \in [0,1]$. But $A_0 \notin \Phi(L_2(0,1))$, since ess $\inf|a(t)| = 0$.

In this example $A \in \Phi(E)$ if and only if $\inf |\gamma_t(A)| \neq 0$ for all $t \in [0,1]$. We next show that an analogous statement is valid in the general case if $BC - CB \in T(E)$ for all $A, B \in \mathcal{V}$.

THEOREM 18.4.   *Suppose that $BC - CB \in T(E)$ for all $B, C \in \mathcal{V}$, and let $\{\gamma_M\}_{M \in \alpha}$ be the extension of a 1-presymbol from $\mathcal{V}$ to the Banach algebra $\mathcal{K}$ generated by the linear manifold $\mathcal{V}$. Then*

$$A \in \Phi(E) \cap \mathcal{K} \iff \inf_{M \in \alpha} |\gamma_M(A)| \neq 0. \qquad (18.9)$$

We need the following lemma:

LEMMA 18.2.   *If the assumptions of Lemma 18.1 are in force and $A \in \mathcal{K} \cap \Phi(E)$, then $\inf |det\ \gamma_M(A)| \neq 0$ for all $M \in \alpha$.*

PROOF.   Suppose that $\inf |det\ \gamma_M(A)| = 0$. Then there exists a sequence $M_n \in \alpha$ such that $det\ \gamma_M(A) \to 0$, and hence a sequence of eigenvalues λ_n of the matrices $\gamma_M(A)$ such that $\lambda_n \to 0$. Then for sufficiently large n, $A - \lambda_n I \in \Phi(E) \cap \mathcal{K}$ although $det\ \gamma_{M_n}(A - \lambda_n I) = 0$, which contradicts Lemma 18.1.

PROOF OF THEOREM 18.4.   The implication $\implies$ follows from Lemma 18.1. Let us show that $(\inf_{M \in \alpha} |\gamma_M(A)| \neq 0) \implies A \in \Phi(E) \cap \mathcal{K}$. First, we prove that under the assumptions of the theorem

$$r(\hat{A}) \leq \sup_{M \in \alpha} |\gamma_M(A)| \qquad (18.10)$$

for all operators $A \in \mathcal{K}$, where $r(\hat{A})$ designates the spectral radius of the element $\hat{A}$. Let $A_n \in \mathcal{K}_0$ be such that $\|A_n - A\| \to 0$. Since $A_n \in \Phi(E)$ if and only if $\gamma_M(A_n) \neq 0$ for all $M \in \alpha$, $r(\hat{A}_n) = \max_{M \in \alpha} |\gamma_M(A_n)|$. But $\hat{A}_n \hat{A} = \hat{A} \hat{A}_n$, and hence

$$r(\hat{A}) \leq r(\hat{A}_n) + r(\hat{A} - \hat{A}_n) \leq \max_{M \in \alpha} |\gamma_M(A_n)| + \|\hat{A} - \hat{A}_n\|. \qquad (18.11)$$

It follows from Corollary 18.3 that $\gamma_M(A_n) \to \gamma_M(A)$ uniformly with respect to M. Consequently, (18.10) follows from (18.11).

We next remark that if $B \in \mathcal{K}_0 \cap \Phi(E)$, then

$$\hat{B}^{-1} - \lambda \hat{I} \in G\widehat{\mathcal{L}(E)} \iff \hat{I} - \lambda \hat{B} \in G\widehat{\mathcal{L}(E)} \iff 1 - \lambda \gamma_M(\hat{B}) \neq 0 \quad \forall M \in \alpha.$$

Therefore,

$$r(\hat{B}^{-1}) \leq \max |\gamma_M^{-1}(B)| = 1/\inf |\gamma_M(B)|. \tag{18.12}$$

Since $\inf |\gamma_M(A)| > 0$,

$$\inf |\gamma_M(A_n)| \geq \delta > 0$$

for sufficiently large n. It then follows from (18.12) that

$$r(\hat{A}_n^{-1}(\hat{A} - \hat{A}_n)) \leq r(\hat{A}_n^{-1})\|\hat{A} - \hat{A}_n\| \leq \delta\|A - A_n\|$$

and hence that $r(\hat{A}_n^{-1}(\hat{A} - A_n)) < 1$ for sufficiently large n. This means that for such values of n, $\hat{A} = \hat{A}_n(I + \hat{A}_n^{-1}(\hat{A} - \hat{A}_n)) \in G\widehat{\mathcal{L}(E)}$.

□

COROLLARY 18.4. *The Banach algebra generated by singular integral operators with continuous coefficients on a compound contour is an algebra with 1-symbol.*

PROOF.   The extension of the presymbol defined in Example 18.2 may be taken as a family of homomorphisms that defines a symbol. In fact, the family of multiplicative functionals defined therein enjoys the property that $\inf |\gamma_M(A)| = |\gamma_{M_0}(A)|$ for some M_0 (which depends on A). Now we can apply Theorem 18.4.

□

Suppose that an n-presymbol is defined on $\mathcal{K}_0$. We shall assume that $\mathcal{K}_0$ and the given presymbol satisfy the following conditions:

– there exist projectors $P_j \in \mathcal{K}_0$, $j = 1, \ldots, n$, such that $P_j P_k = \delta_{jk} P_j$, $j, k = 1, \ldots, n$, and $\sum_{j=1}^{n} P_j = I$;

– the subalgebras $\hat{P}_j \hat{\mathcal{K}} \hat{P}_j$, $j = 1, \ldots, n$ are commutative;

– $\mathcal{K}$ is inverse-closed in $\widehat{\mathcal{L}(E)}$;

– there exists a constant β such that $|a_{pq}(M)| \leq \beta\|\hat{A}\|$ for all $A \in \mathcal{K}_0$ and all $M \in \alpha$ (where $\gamma_M(A) = [a_{pq}(M)]$);

– $\gamma_M(P_j) = [\delta_{pq}\varepsilon_p]$, where $\varepsilon_m = 1$ and $\varepsilon_p = 0$ for $p \neq m$.

THEOREM 18.5. *Let $\{\gamma_M\}_{M \in \alpha}$ be the extension of the n-presymbol to $\mathcal{K}$. If $A \in \mathcal{K}$, then*

$$A \in \Phi(E) \iff \inf_{M \in \alpha} |\det \gamma_M(A)| \neq 0. \tag{18.13}$$

We need the following lemma:

LEMMA 18.3. *Under the assumption of Theorem* 18.5, *let* $A_0 \in \mathcal{K}$ *and* $\gamma_M(A_0) = [a_{jk}(M)]$. *Then there exists a* $\delta > 0$ *such that if* $|a_{jk}(M)| \leq \delta$ $\forall M \in \alpha$, *then* $A = I - A_0 \in \Phi(E)$.

PROOF. We proceed by induction on n. For $n = 1$ the assertion follows from Theorem 18.4. For the general case we need some preparation. Let $P_1 = P$, $I - P_1 = Q$, let $\mathcal{K}'$ ($\mathcal{K}'_0$) be the algebra of operators of the form $A' = PAP + \alpha Q$ with $\alpha \in \mathbb{C}$ and $A \in \mathcal{K}$ (respectively, $A \in \mathcal{K}_0$), and let $\mathcal{K}''$ ($\mathcal{K}''_0$) be the algebra of operators of the form $A'' = QAQ + \beta P$ with $\beta \in \mathbb{C}$ and $A \in \mathcal{K}$ (respectively, $A \in \mathcal{K}_0$). If $\gamma_M(A) = [b_{jk}(M)]_{j,k=1}^n$, we set $\gamma_M''(A'') = [b_{jk}]_{j,k=2}^n$, $f_M''(A'') = \beta$, $\gamma_M'(A') = b_{11}(M)$, and $f_M'(A') = \alpha$. It is readily verified that the family of homomorphisms $\{\gamma_M'\} \cup \{f_M'\}$ (respectively, $\{\gamma_M''\} \cup \{f_M''\}$) defines an $(n-1)$-symbol (respectively, a 1-symbol) on $\mathcal{K}'_0$ (respectively, $\mathcal{K}''_0$). For sufficiently small δ, $\inf |1 - a_{11}(M)| > 0$ and hence, by Theorem 18.4, $PAP + Q \in \Phi(E)$. It follows from the conditions listed just before Theorem 18.5 that there exists an operator $B \in \mathcal{K}$ such that $PBP + Q$ is a regularizer for $PAP + Q$. A straightforward calculation shows that

$$A = (PAP + Q)(I + QAP)[P + Q(A - APBPA)Q](I + PBPAQ) + T, \qquad (18.14)$$

where $T \in \mathcal{T}(E)$. The operators $I + QAP$ and $I + PBPAQ$ are invertible: $(I + QAP)^{-1} = I - QAP$, and, in general, $(I + PXQ)^{-1} = I - PXQ$. We already know that $PAP + Q \in \Phi(E)$. The operator $A'' = P + Q(A - APBPA)Q$ belongs to $\Phi(E)$ by the inductive hypothesis: in fact, $f''(A) = 1$ and the entries of the matrix $\gamma_M''(I - A'')$ can be made sufficiently small by conveniently choosing the initial $\delta > 0$. Hence, $A \in \Phi(E)$, as claimed.

□

PROOF OF THEOREM 18.5. The implication $\Longrightarrow$ follows from Lemma 18.2. Let us prove the converse. Let $A \in \mathcal{K}$, $\inf_M |det \ \gamma_M(A)| > 0$. Let $A_n \in \mathcal{K}_0$ such that $\|A - A_n\| \to 0$. Since $A_n \in \Phi(E)$ for sufficiently large n and since $(\gamma_M(A))^{-1} \gamma_M(A - A_n)$ converges uniformly to the null matrix we have, in view of Lemma 18.3, that

$$\hat{I} - \hat{A}_n^{-1}(\hat{A} - \hat{A}_n) \in G\widehat{\mathcal{L}(E)} \Longrightarrow \hat{A} \in G\widehat{\mathcal{L}(E)} \Longrightarrow A \in \Phi(E). \qquad \square$$

Remark 18.2. In Theorem 18.5 the requirement that the algebra $\mathcal{K}$ be inverse-closed may be replaced by the requirements that the subalgebra $\hat{P}_j\hat{\mathcal{K}}\hat{P}_j$ be inverse-closed in $\hat{P}_j\widehat{\mathcal{L}(E)}\hat{P}_j$, and that the set of invertible elements of $\hat{P}_j\hat{\mathcal{K}}\hat{P}_j$ be dense in this algebra for all $j = 1,\ldots,n$. The fact that these last two conditions imply the first can readily be established by induction on n with the help of equality (18.14).

THEOREM 18.6. *The algebra of Example* 18.1, *generated by singular integral operators with piecewise-continuous coefficients, is an algebra with a $\mathcal{R}$-symbol (which is the extension of the presymbol of Example* 18.1).

PROOF. We show that the conditions of Theorem 18.5 and Remark 18.2 are fulfilled. The role of the projectors P_j is played by $P = \frac{1}{2}(I + S_\Gamma)$ and $Q = I - P$. The fact that the conditions formulated in Remark 18.2 are fulfilled follows from Lemma 16.5, Theorem 16.5, and its proof. It remains to check that

$$|a_{jk}(M)| \leq \beta\|\hat{A}\| \tag{18.15}$$

where β does not depend on A and M. Using theorem 18.3, we can take $R_{11} = P$ and $R_{22} = Q$ and deduce that inequality (18.15) holds for $a_{jj}(M)$. Unfortunately, in algebra $\mathcal{K}_0$ there are no operators that may serve as R_{jk} for $j \neq k$ simultaneously for all $M = (t, \mu)$. Some extra work is needed to circumvent this difficulty. We need an orientation reversing one-to-one mapping ω of the contour Γ onto itself which enjoys the following properties: $\omega(\omega(t)) = t$; $\omega(t_0) = t_0$, where t_0 is a prescribed point; $\omega'(t)$ satisfies Hölder's condition on Γ, and $|\omega'(t)| \leq c$ (where c does not depend on the point t_0 that ω must keep fixed). If $\Gamma = \Gamma_0$ we take $\omega(t) = t_0^2/t$. If Γ is a simple closed Lyapunov curve we set $\omega(t) = \beta(\omega_1(\alpha(t)))$, where $\beta\colon \Gamma_0 \to \Gamma$, $\alpha = \beta^{-1}$, $\omega_1(z) = \alpha(t_0)^2/z$, and $\beta'(z)$ satisfies Hölder's condition on Γ_0. Let V be the operator defined by the rule $(V\varphi)(t) = \varphi(\omega(t))$. Then V is a bounded linear operator from $L_p(\Gamma, \rho)$ into $L_p(\Gamma, \rho_1)$ and also from $L_p(\Gamma, \rho_1)$ into $L_p(\Gamma, \rho)$, where $\rho_1 = \prod |t - \omega(t_k)|^{\beta_k}$. It is readily checked that in either case

$$\|V\| \leq \max |\omega'(t)|^\delta,$$

where $\delta = (1 + \sum_k |\beta_k|)p^{-1}$, and that

$$VaV = a(\omega(t))I \quad \text{and} \quad VSV = -S + T \tag{18.16}$$

with $T \in \mathcal{T}(E)$. Let $A \in \mathcal{K}_0, \gamma_{t,\mu}(A) = [a_{jk}(t,\mu)]$, and $B = PAQ$. Then

$$\gamma_{t,\mu}(A) = \begin{bmatrix} 0 & a_{12}(t,\mu) \\ 0 & 0 \end{bmatrix}. \tag{18.17}$$

Upon replacing Q by $I - P$ into the right-hand side of the equality $A = \sum A_{j1}A_{j2}\cdots A_{jr}$ (where $A_{jk} = a_{jk}P + b_{jk}Q$) we can write the operator B as

$$B = \sum_j Pc_{j1}Pc_{j2}P\cdots Pc_{jr'}Pc_jQ \tag{18.18}$$

where $c_{jk}, c_j \in PC(\Gamma)$. Now consider the operator $X = VBV$ in $L_p(\Gamma,\rho)$. By (18.16) and (18.8),

$$X = \sum_j Qd_{j1}Qd_{j2}Q\cdots Qd_{jr'}Qd_jP + T', \tag{18.19}$$

where $T' \in \mathcal{T}(E)$, $d_{jm}(t) = c_{jm}(\omega(t))$, and $d_j(t) = c_j(\omega(t))$. It can be readily shown that

$$\gamma_{t,\mu}(X) = \begin{bmatrix} 0 & 0 \\ x_{21}(t,\mu) & 0 \end{bmatrix}.$$

Since $d_{jm}(t_0 \pm 0) = c_{jm}(t_0 \mp 0)$ and $d_j(t \pm 0) = c_j(t_0 \mp 0)$, it follows that

$$x_{21}(t_0,\mu) = -a_{12}(t_0,\mu). \tag{18.20}$$

Set $Z = BX$. Then

$$\gamma_{t_0,\mu}(Z) = \begin{bmatrix} -a_{12}^2(t_0,\mu) & 0 \\ 0 & 0 \end{bmatrix},$$

and, by a previous result (see Theorem 18.3), $|a_{12}^2(t_0,\mu)| \leq \|\hat{P}\|^2\|\hat{Z}\|$. But $\|\hat{Z}\| \leq \|\hat{A}\|^2\|V\|^2\|P\|^2\|Q\|^2$, whence $|a_{12}(t,\mu)| \leq \beta\|\hat{A}\|$ (using the fact that $\|V\|$ does not depend on t_0).

$\square$

19. AN ALGEBRA OF SINGULAR INTEGRAL OPERATORS WITH CARLEMAN SHIFT

Let Γ be an oriented Lyapunov contour in the complex plane, $E = L_p(\Gamma)$ (with $1 < p < \infty$), Σ a subalgebra of $L_\infty(\Gamma)$, $\alpha : \Gamma \to \Gamma$ a homeomorphism such that $\alpha(t) \not\equiv t$, $\alpha(\alpha(t)) \equiv t$, and the derivative $\alpha'(t)$ satisfies Hölder's condition on Γ. We let $\mathcal{K} = \mathcal{K}(\Sigma)$ denote the Banach algebra generated by the operator S, the compact operators, and the operators of multiplication by functions $h \in \Sigma$. Let $\tilde{\mathcal{K}} = \tilde{\mathcal{K}}(\Sigma)$ denote the Banach subalgebra of $\mathcal{L}(E)$ generated by $\mathcal{K}$ and the Carleman shift operator $(V\varphi)(t) = \varphi(\alpha(t))$.

It is readily verified that if the map α preserves (reverses) orientation, then $SVS = V + T$ (respectively, $SVS = -V + T_1$), where $T \in \mathcal{T}(E)$ (respectively, $T_1 \in \mathcal{T}(E)$). Moreover, if $(H\varphi)(t) = h(t)\varphi(t)$ is a multiplication operator, then $(VHV\varphi)(t) = h(\alpha(t))\varphi(t)$.

LEMMA 19.1. *Suppose that α is orientation-preserving. Then every operator $A \in \tilde{\mathcal{K}}$ can be expressed as a sum $A = A_1 + A_2 V$, where $A_1, A_2 \in \mathcal{K}$ and are uniquely determined modulo compact operators.*

PROOF. It obviously suffices to show that the set $\mathcal{N}$ of operators of the form

$$A_1 + A_2 V \tag{19.1}$$

where A_1 and A_2 runs through $\mathcal{K}$, is a Banach algebra, and that from the equality $A_1 + A_2 V = 0$, with $A_1, A_2 \in \mathcal{K}$, it follows that $A_1, A_2 \in \mathcal{T}(E)$. It is readily checked that the sum and product of operators of the form (19.1) are again of the same form. Let us show that $\mathcal{N}$ is closed. Suppose that the sequence $A_{1,n} + A_{2,n} V$ converges to $A = A_1 + A_2 V$ in the uniform norm, and let $h(t) = \alpha(t) - t$ and $(H\varphi)(t) = h(t)\varphi(t)$. Since $H^{-1}(A_{1,n} + A_{2,n}V)H = A_{1,n} - A_{2,n}V + T_n$ with $T_n \in \mathcal{T}(E)$, it follows that $2\hat{A}_{1,n})\hat{A} + \hat{H}^{-1}\hat{A}\hat{H} = 2\hat{A}_1 \in \hat{\mathcal{K}}$. Similarly, $\hat{A}_{2,n})\hat{A}_2 \in \hat{\mathcal{K}}$. Therefore, $\hat{A} = \hat{A}_1 + \hat{A}_2\hat{V}$, and hence $A = A_1 + A_2 V$ with $A_1, A_2 \in \mathcal{K}$.

Now let $A_1 + A_2 V = 0$. Then $\hat{A}_1 + \hat{A}_2\hat{V} = 0$ and $\hat{A}_1 - \hat{A}_2\hat{V} = \hat{H}^{-1}(\hat{A}_1 + \hat{A}_2\hat{V})\hat{H} = 0$, i.e., $A_1, A_2 \in \mathcal{T}(E)$. □

LEMMA 19.2. *Let $A, B, C \in \mathcal{L}(E)$ and $C^2 = I$. Then*

$$\begin{bmatrix} I & C \\ I & -C \end{bmatrix} \begin{bmatrix} A & B \\ CBC & CAC \end{bmatrix} \begin{bmatrix} I & I \\ C & -C \end{bmatrix} = 2 \begin{bmatrix} A + BC & 0 \\ 0 & A - BC \end{bmatrix}. \qquad (19.2)$$

PROOF. Straightforward verification.

$\square$

THEOREM 19.1. *Let $A, B \in \mathcal{K}$ and suppose that the map α preserves the orientation of Γ. Then*

$$A + BV \in \Phi(E) \Longleftrightarrow \begin{bmatrix} A & B \\ VBV & VAV \end{bmatrix} \in \Phi(E^2). \qquad (19.3)$$

Moreover, if $A + BV \in \Phi(E)$, then

$$Ind(A + BV) = \frac{1}{2} Ind \begin{bmatrix} A & B \\ VBV & VAV \end{bmatrix}. \qquad (19.4)$$

PROOF. We use equality (19.2) for the operators A, B, and $C = V$. We remark that $H^{-1}(A - BV)H = A + BV + T$ with $T \in \mathcal{T}(E)$. Consequently, $A + BV \in \Phi(E) \Longleftrightarrow A - BV \in \Phi(E)$ and $Ind(A + BV) = Ind(A - BV)$. Since the outer factors in the left hand side of equality (19.2), written for A, B and $C = V$, are invertible operators (their product is equal to $2 \begin{bmatrix} I & 0 \\ 0 & I \end{bmatrix}$), the assertions of the theorem are consequences of (19.2).

$\square$

Remark 19.1. The matrix operator in (19.3) and (19.4) is a singular integral operator with matrix coefficients (without shift).

COROLLARY 19.1. *Let Γ be a simple closed Lyapunov contour, let $A = aI + bS$ and $B = cI + dS$ where $a, b, c, d \in C(\Gamma)$, and let $R = A + BV$. Suppose that α preserves orientation. Then $R \in \Phi(E) \Longleftrightarrow$*

$$det \begin{bmatrix} a(t) + b(t) & c(t) + d(t) \\ c(\alpha(t)) + d(\alpha(t)) & a(\alpha(t) + b(\alpha(t)) \end{bmatrix} \begin{bmatrix} a(t) - b(t) & c(t) - d(t) \\ c(\alpha(t)) + d(\alpha(t)) & a(\alpha(t)) + b(\alpha(t)) \end{bmatrix} \neq 0 \qquad (19.5)$$

If $R \in \Phi(E)$, then

$$Ind\ R = \frac{1}{2}Ind\ det(f_1(t)f_2(t)^{-1})$$

where f_1 and f_2 are the matrices appearing in (19.5) in the indicated order.

□

COROLLARY 19.2.   *Let $\Sigma = PC(\Gamma)$ and let $\{\gamma_M\}$ be a family of homomorphisms which defines a 2-symbol on the algebra $\mathcal{K}$. Suppose that α preserves orientation. Then the family of homomorphisms*

$$\tilde{\gamma}_M(A) = \begin{bmatrix} \gamma_M(A_1) & \gamma_M(A_2) \\ \gamma_M(VA_2V) & \gamma_M(VA_1V) \end{bmatrix} \tag{19.6}$$

defines a 4-symbol on the algebra $\tilde{\mathcal{K}}$.

□

Now suppose that α reverses the orientation of Γ. Then Theorem 19.1 remains valid when Γ is a closed Lyapunov contour and $\Sigma = C(\Gamma)$, and the proof is analogous. One has only to replace the operator H by S and use the properties: $aS - SaI \in T(E)$ and $SVS + V \in T(E)$.

When $\Sigma = PC(\Gamma)$, Theorem 19.2 is not applicable to the case where α reverses orientation. To show this it suffices to exhibit two singular integral operators A and B such that $A + BV \in \Phi(E)$ but $A - BV \notin \Phi(E)$. Then it follows from equality (19.2) that the corresponding matrix operator is not Fredholm.

Example 19.1.   Take $E = L_2(\mathbb{R})$, $(V\varphi)(t) = \varphi(-t)$, $A = I$, and $B = aS$, where $a(t) = 0$ for $t < 0$ and $a(t) = i$ for $t > 0$. Let us show that $A - BV \in \Phi(E)$, but $A + BV \notin \Phi(E)$. To this end we consider the operator $C = A + \lambda BV$ and the isomorphism $v : L_2(\mathbb{R}) \to L_2^2(\mathbb{R}_+)$ defined by the rule: $(V\varphi)(t) = (\varphi(t), \varphi(-t))$, $t > 0$. Then

$$vCv^{-1} = \begin{bmatrix} I & 0 \\ 0 & I \end{bmatrix} + \lambda \begin{bmatrix} iI & 0 \\ 0 & 0 \end{bmatrix} \begin{bmatrix} S_0 & -R \\ R & -S_0 \end{bmatrix} \begin{bmatrix} 0 & I \\ I & 0 \end{bmatrix}$$

$$= \begin{bmatrix} I - i\lambda R & \lambda i S_0 \\ 0 & I \end{bmatrix}, \tag{19.7}$$

where S_0 and R are defined by formulas (17.4).

It follows from (19.7) that $C \in \Phi(E)$ if and only if $I - i\lambda R \in \Phi(L_2(\mathbb{R}_+))$. The operator $I - i\lambda R$ belongs to the algebra $\mathcal{K}_0$ generated by S_0 in $L_2(\mathbb{R}_+)$. By Theorem 13.7,

$$I - i\lambda R \in G\mathcal{L}(L_2(\mathbb{R}_+)) \iff (1 - i\lambda\sqrt{z^2 - 1} \neq 0 \quad \forall\, z \in [-1, 1]) \iff \lambda \notin [1, \infty) \; ;$$

here the branch of the square root $\sqrt{z^2 - 1}$ is chosen so that $\sqrt{-1} = i$. Therefore, $I - i\lambda R$ is invertible for $\lambda = -1$ and noninvertible for $\lambda = 1$. The operator iR is self-adjoint and, as previously shown, its spectrum coincides with the segment $[0, 1]$. For this operator the boundary point $\lambda = 1$ of the spectrum is not a Φ-point, and hence $I - iR \notin \Phi(L_2(\mathbb{R}_+))$. Returning to the operator C, we conclude that $A + BV \notin \Phi(E)$, whereas $A - BV \in \Phi(E)$, as claimed.

The fact that we were dealing with the real line rather than a closed contour is immaterial since the former can always be mapped onto a closed contour (say, Γ_0) and in this way we can build the corresponding example in $L_2(\Gamma_0)$. Therefore, the rule (19.6) does not define a symbol on $\tilde{\mathcal{K}}$ in the case where α reverses the orientation of the contour.

We next indicate how one can construct a symbol in this case.

LEMMA 19.3. *Let Γ be a simple, oriented, closed Lyapunov contour and let $\alpha:\Gamma \to \Gamma$ be an orientation-reversing homeomorphism such that $\alpha(\alpha(t)) \equiv t$ and $\alpha'(t)$ satisfies Hölder's condition on Γ. Then there exists an invertible mapping ξ of the unit circle Γ_0 onto Γ with the following properties:*

$$\xi'(z) \in H(\Gamma_0) \, , \; \xi'(z) \neq 0 \quad for\ all \;\; z \in \Gamma_0 \, , \tag{19.8}$$

and

$$(\xi^{-1}\alpha\xi)(z) = z^{-1} \, . \tag{19.9}$$

(Here $H(\Gamma_0)$ denotes the set of functions which satisfy Hölder's condition on Γ_0.)

PROOF. It is readily checked that the involution α has exactly two fixed points on Γ, which we denote by t_0 and τ_0. Since Γ is a Lyapunov contour, there is a mapping $\eta:\Gamma_0 \to \Gamma$ such that $\eta'(z) \in H(\Gamma_0)$ and $\eta'(z) \neq 0$ for all $z \in \Gamma_0$. Moreover, one can choose η

so that $\eta(1) = t_0$. Set $\lambda = \eta^{-1}\alpha\eta$. It is easily seen that $\lambda{:}\Gamma_0 \to \Gamma_0$ reverses the orientation of the circle. Moreover, $\lambda(\lambda(z)) = z$ $(|z| = 1)$ and $\lambda(\zeta_0) = \zeta_0$, where $\zeta_0 = \eta^{-1}(t_0)$. The points 1 and ζ_0 divide Γ_0 into two arcs Γ_1, Γ_2 such that λ maps Γ_1 into Γ_2 and Γ_2 into Γ_1. Let $\mu{:}\Gamma_1 \to \Gamma_2$ be a mapping which satisfies the conditions $\mu(\zeta_0) = -1$, $\mu(1) = 1$, $\mu(z) \neq 0$ for $z \in \Gamma_1$, and $\mu'(z) \in H(\Gamma_1)$. We extend μ to the entire circle Γ_0 setting $\mu(z) = 1/\mu(\lambda(z))$ for $z \in \Gamma_2$. It is readily verified that $\mu(z) = 1/\mu(\lambda(z))$ for every point $z \in \Gamma$. Set

$$\mu'(\zeta_0 + 0) = \lim_{\substack{\zeta > \zeta_0 \\ \zeta \in \Gamma_1}} \mu'(\zeta) \quad \text{and} \quad \mu'(\zeta_0 - 0) = \lim_{\substack{\zeta > \zeta_0 \\ \zeta \in \Gamma_2}} \mu'(\zeta) .$$

Since

$$\mu'(\zeta_0 - 0) = -\mu'(\zeta_0 + 0)\lambda'(\zeta_0)[\mu(\zeta_0)]^{-2} ,$$

$\lambda'(\zeta_0) = -1$, and $\mu(\zeta_0) = -1$, it follows that $\mu'(\zeta_0 - 0) = \mu'(\zeta_0 + 0)$. It is similarly checked that $\mu'(1 - 0) = \mu'(1 + 0)$. Therefore, $\mu' \in H(\Gamma_0)$. Finally, it is readily verified that the mapping $\xi = \eta\mu^{-1}$ enjoys the desired properties.

Let $\Sigma \subset L_\infty(\Gamma)$ be an algebra of functions $f : \Gamma \to \mathbb{C}$, as above, and let $\tilde{\mathcal{K}}$ denote the algebra generated by the operators $H = hI$ with $h \in \Sigma$, S_Γ, $T \in \mathcal{T}(E)$, and V, where $(V\varphi)(t) = \varphi(\alpha(t))$. For simplicity we shall consider the case $E = L_2(\Gamma)$.

Let $B : L_2(\Gamma) \to L_2(\Gamma_0)$ be the isomorphism defined by the formula $(B\varphi)(z) = \varphi(\xi(z))$. It is easily seen that $(BHB^{-1}\varphi)(z) = h(\xi(z))I$, $BS_\Gamma B^{-1} = S_0 + T$, where $T \in \mathcal{T}(L_2(\Gamma_0))$, and $(BVB^{-1}\varphi)(z) = \varphi(1/z)$. Hence, the general case reduces to the case $\Gamma = \Gamma_0$ and $\alpha(z) = 1/z$.

Let $c : L_2(\Gamma_0) \to L_2(\mathbb{R})$ be the isomorphism defined by the rule

$$(c\varphi)(x) = \frac{2i}{i+x}\varphi\left(\frac{i-x}{i+x}\right) , \quad x \in \mathbb{R} .$$

Then

$$(c^{-1}f)(z) = \frac{1}{z+1}f\left(i\frac{1-z}{1+z}\right) , \quad z \in \Gamma_0 .$$

Set $(W\varphi)(z) = \varphi(1/z)$. Then $(cWc^{-1}f)(t) = (i-t)(i+t)^{-1}f(-t)$ and $cS_0c^{-1} = S_\mathbb{R}$. This shows that $\tilde{\mathcal{K}}$ is isomorphic to the algebra generated by the singular integral operators with the shift $(u\varphi)(x) = \varphi(-x)$ and coefficients in a certain subalgebra of $L_\infty(\mathbb{R})$.

Finally, we consider the isomorphism $\mathcal{D}: L_2(\mathbb{R}) \rangle L_2^2(\mathbb{R}_+)$ defined by the rule $(\mathcal{D}\varphi)(x) = (\varphi(x), \varphi(-x))$. We already know that

$$\mathcal{D}H\mathcal{D}^{-1} = \begin{bmatrix} H_1 & 0 \\ 0 & H_2 \end{bmatrix}, \quad \mathcal{D}S_\mathbb{R}\mathcal{D}^{-1} = \begin{bmatrix} S & -R \\ R & -S \end{bmatrix}, \quad \mathcal{D}u\mathcal{D}^{-1} = \begin{bmatrix} 0 & I \\ I & 0 \end{bmatrix},$$

where H is an arbitrary multiplication operator in $L_2(\mathbb{R})$.

Let $M = \mathcal{D}cB$, $\Sigma = PC(\Gamma)$, and $A \in \tilde{\mathcal{K}}$. Then

$$A \in \Phi(L_2(\Gamma)) \Longleftrightarrow MAM^{-1} \in \Phi(L_2^2(\mathbb{R}_+)) .$$

The operators MAM^{-1} belong to the algebra generated by the singular integral operators with piecewise-continuous 2×2 matrix coefficients. Consequently, a 4-symbol can be introduced on $\tilde{\mathcal{K}}$.

Usually a shift $\alpha{:}\Gamma \to \Gamma$ is called a *Carleman shift* [53] if one of its iterations is the identity mapping. If Γ is a simple closed Lyapunov contour, then every orientation-reversing Carleman shift α has order 2, i.e., its second iteration is the identity mapping of Γ. If, however, α preserves orientation, then it can have arbitrary order. For example, the rotation of the circle by an angle of $2\pi/n$ has order n. On algebras generated by singular integral operators with piecewise-continuous coefficients and Carleman shift operators of order n one can introduce a $2n$-symbol (see [53]).

We shall see in Sec.27 that algebras that contain a non-Carleman shift and multiplication operators admit no n-symbols whatever the value of n.

CHAPTER VI

SUFFICIENT FAMILIES OF n-DIMENSIONAL REPRESENTATIONS OF BANACH ALGEBRAS

In this chapter we construct a matrix analog of Gelfand's transformation, we investigate its properties, and we give examples.

20. THE STANDARD IDENTITY ON $\mathbb{C}^{n \times n}$

Let $\mathcal{K}$ be an algebra, $a_1, \ldots, a_m \in \mathcal{K}$, and let F_m be the *standard polynomial*

$$F_m(a_1, \ldots, a_m) = \sum_{\sigma \in S_m} sgn\ \sigma a_{\sigma(1)} \cdots a_{\sigma(m)} \ , \tag{20.1}$$

where σ runs through the symmetric group S_m.

We say that the algebra $\mathcal{K}$ *satisfies a standard identity (of order m)* if $F_m(a_1, \ldots, a_m) = 0$ for every collection $a_1, \ldots, a_m$ of elements of $\mathcal{K}$.

For instance, $F_2(a, b) = ab - ba$, so that algebra $\mathcal{K}$ satisfies the standard identity of order 2 if and only if it is commutative. In this respect the class of algebras satisfying standard identities is a generalization of the class of commutative algebras.

The standard polynomial can be also defined recursively:

$$F_1(a) = a$$

and

$$F_m(a_1, \ldots, a_m) = \sum_{i=1}^{m} (-1)^{i-1} a_i F_{m-1}(a_1, \ldots, \hat{a}_i, \ldots, a_m) \tag{20.1}$$

$(m > 1)$, where $\wedge$ indicates that the corresponding element is deleted.

THEOREM 20.1. (Amitsur-Levitzki). *The algebra $\mathbb{C}^{n \times n}$ of complex $n \times n$ matrices satisfies the standard identity of order $2n$.*

We give two different proofs of this important and interesting theorem.

FIRST PROOF. In view of the multi-linearity of the polynomial F_m, it suffices to prove that the equality

$$F_{2n}(a_1, a_2, \ldots, a_{2n}) = 0 , \quad a_j \in \mathbb{C}^{n \times n} , \tag{20.2}$$

holds for the matrix units e_{pq} (i.e., the matrices with entries $(e_{pq})_{k\ell} = \delta_{pk}\delta_{q\ell}$, where δ_{pk} is the Kronecker symbol. For brevity we shall write e_m instead of e_{mm}. Notice that $e_{pq}e_{r\ell} = \delta_{qr}e_{p\ell}$.

We proceed by induction. For $m = 1$, equality (20.2) is obvious. Suppose that it holds for $m = n - 1$. We prove that it holds for $m = n$. Let

$$\mathcal{E} = \{e_{i_1 j_1}, e_{i_2 j_2}, \ldots, e_{i_{2n} j_{2n}}\}$$

be a collection of matrix units. For any such collection we let $f_{\mathcal{E}}(k)$ denote the number of times the index k $(1 \leq k \leq n)$ appears in the matrices e_{ij} that form $\mathcal{E}$ (we consider that the matrix e_{kk} has two k indices). Then obviously,

$$1 \leq k \leq n \quad \text{and} \quad \sum_{k=1}^{n} f_{\mathcal{E}}(k) = 4n . \tag{20.3}$$

We prove the equality (20.2) separately for the various possible choices of collections $\mathcal{E}$.

(1) Suppose there exist an index k such that $f_{\mathcal{E}}(k) = 0$. Then we can regard the matrix units in $\mathcal{E}$ as matrix units of order $n - 1$, and (20.2) is seen to be a consequence of (20.1). A similar argument works when $e_k \in \mathcal{E}$ and $f_{\mathcal{E}}(k) = 2$.

(2) Suppose $e_k \in \mathcal{E}$ and $f_{\mathcal{E}}(k) = 3$. In this case the nonzero terms in $F_{2n}(e_{i_1 j_1}, \ldots, e_{i_{2n} j_{2n}})$ are necessarily of one of the following types: $e_k e_{kp}$ ... or $e_{pk}e_k$. The sum of the terms of first type equals

$$\pm e_k e_{kp} F_{2n-2}(e_{i_1 j_1}, \ldots, \hat{e}_k, \ldots, \hat{e}_{kp}, \ldots, e_{i_{2n} j_{2n}}) .$$

Since the index k does not appear in $F_{2n-2}(\ldots)$, it follows that $F_{2n}(\ldots) = 0$. Similarly, the sum of the terms of second type is equal to zero.

(3) Suppose $e_k \in \mathcal{E}$ and $f_{\mathcal{E}}(k) = 4$. Here the nonnull terms are of three types: $e_k e_{kp} a e_{qk}$; $e_{kp} a e_{qk} e_k$, and $a e_{qk} e_k e_{kp} b$. Since $e_{kp} a e_{qk}$ has an odd number of factors, the terms $e_k e_{kp} a e_{qk} = \pm e_k$ and $e_{kp} a e_{qk} e_k = \mp e_k$ have opposite signs and cancel one another. The sign with which the monomial

$$x_{\sigma(1)} \cdots x_{\sigma(i)} e_{qk} e_k e_{kp} x_{\sigma(i+1)} \cdots x_{\sigma(2n-3)}$$

appears in the sum $F_{2n}(e_{qp}, e_k, e_{kp}, x_1, x_2, \ldots, x_{2n-1})$ coincides with the sign of the monomial

$$x_{\sigma(1)} \cdots x_{\sigma(i)} e_{qp} x_{\sigma(i+1)} \cdots x_{\sigma(2n-3)}$$

in the sum $F_{2n-2}(e_{qp}, x_1, x_2, \ldots, x_{2n-3})$, and the two monomials coincide. Consequently, the sum of the terms of the third type is equal to $\pm F_{2n-2}(e_{qp}, \ldots, x_{2n-3}) = 0$ by the inductive hypothesis.

(4) Suppose $e_k \in \mathcal{E}$ for all $k = 1, 2, \ldots, n$. Since $\Sigma f_{\mathcal{E}}(k) = 4n$, there is an index k such that $f_{\mathcal{E}}(k) \leq 4$. But (20.2) holds for such a collection $\mathcal{E}$ by the foregoing argument.

The last case analyzed permits us to alter the course of the proof. Specifically, we shall assume that (20.2) holds for any collection $\mathcal{E}$ which contains $r + 1$ idempotent elements $e_1, \ldots, e_{r+1}$ and prove that it holds for any collection $\mathcal{E}$ which contains $e_1, \ldots, e_r$, where $0 \leq r \leq n - 1$. By the foregoing discussion, we may assume that if $\mathcal{E} = \{x_1, \ldots, x_{2n}\} \subset \mathbb{C}^{n \times n}$ and there exists an automorphism π of $\mathbb{C}^{n \times n}$ which takes $(x_1, \ldots, x_{2n})$ into $(e_1, e_2, \ldots, e_{r+1}, e_{ij}, \ldots, e_{pq})$, then

$$F_{2n}(x_1, \ldots, x_{2n}) = 0 . \tag{$\widetilde{20.2}$}$$

(5) Suppose $e_1, \ldots, e_r \in \mathcal{E}$ and there is a matrix $e_{ij} \in \mathcal{E}$ such that $i > r$, $j > r$, $i \neq j$. In this case $(e_1, e_2, \ldots, e_r, e_i + e_{ij})$ is taken by π into $(e_1, e_2, \ldots, e_r, e_i)$, and by $\widetilde{(20.2)}$

$$F_{2n}(e_1, e_2, \ldots, e_r, e_{ij} + e_i, \ldots) = 0 = F_{2n}(e_1, e_2, \ldots, e_r, e_i, \ldots) ,$$

which in turn implies that

$$F_{2n}(e_1, e_2, \ldots, e_r, e_{ij}, \ldots) = 0 .$$

(6) Suppose $e_1, \ldots, e_r \in \mathcal{E}$ and there is no $\mathbf{e}_{ij} \in \mathcal{E}$ such that $i > r$ and $j > r$. This means that if $e_{ij} \in \mathcal{E}\backslash\{e_1, \ldots, e_r\}$, then $i \le r$ or $j \le r$. In particular, $r > 1$. We may assume that $f_\mathcal{E}(k) \ge 1$ for all $k = 1, \ldots, n$ (see Case 1).Now notice that if the sum $F_{2n}(e_{i_1 j_1}, \ldots, e_{i_{2n} j_{2n}})$ contains nonnull elements, then either $f_\mathcal{E}(k)$ is even for all k, or $f_\mathcal{E}(k)$ is odd for exactly two values k_1 and k_2 of k, and then the indices of the nonnull terms begin (end) with k_1 (respectively, k_2); this is a consequence of the rule for multiplying matrix units. By case (4), we may assume that $f_\mathcal{E}(i) > 4$ for all $i = 1, 2, \ldots, r$. This implies, in view of (20.3), that there is an index $i > r$ such that $f_\mathcal{E}(i) \le 3$. Actually, we can even show that there is an index $i > r$ such that $f_\mathcal{E}(i) \le 2$. In fact, assuming the contrary, we would have $f_\mathcal{E}(1) > 4, \ldots, f_\mathcal{E}(r) > 4, f_\mathcal{E}(r + 1) \ge 3, \ldots, f_\mathcal{E}(n) \ge 3$. Since $\Sigma f_\mathcal{E}(k) = 4n$, $f_\mathcal{E}(k)$ must be odd for some values of k (by the foregoing discussion, there are exactly two such values). But this contradicts the equality $\Sigma f_\mathcal{E}(k) = 4n$, because $r > 1$.

Thus, it remains to examine the following case: $e_1, \ldots, e_r \in \mathcal{E}$, there is a $j > r$ such that $f_\mathcal{E}(j) \le 2$, and for an index $i \le r$ either$e_{ij} \in \mathcal{E}$ or $e_{ji} \in \mathcal{E}$. If $e_{ij} \in \mathcal{E}$, then

$$(e_1, \ldots, e_{i-1}, e_i - e_{ij}, e_{i+1}, \ldots, e_r, e_{ij} + e_j, \ldots)\overset{\pi}{\sim}(e_1, \ldots, e_{r+1}, \ldots) ,$$

and, by (20.3),

$$F_{2n}(\ldots e_i \ldots e_{ij} \ldots) - F_{2n}(\ldots e_{ij} \ldots e_{ij}) +$$
$$F_{2n}(\ldots e_i \ldots e_j \ldots) - F_{2n}(\ldots e_{ij} \ldots e_j \ldots) = 0 . \qquad (20.4)$$

The second term in (20.4) vanishes because e_{ij} appears twice, the second term vanishes because it contains $r + 1$ elements e_k, and the fourth term vanishes because $f_\mathcal{E}(j) \le 4$. We conclude that $F_{2n}(\ldots e_i \ldots e_{ij} \ldots) = 0$.

Finally, if $e_{ji} \in \mathcal{E}$ we set $x_i = e_i - e_{ji}$, $x_{r+1} = e_{ji} + e_j$, and argue in the same fashion.

SECOND PROOF. Define ψ_{2n} by

$$\psi_{2n}(a_1, \ldots, a_{2n}) = \sum_{\sigma \in S_{2n}} sgn\ \sigma[a_{\sigma(1)}, a_{\sigma(2)}] \ldots [a_{\sigma(2n-1)}, a_{\sigma(2n)}] ,$$

where $[a, b] = ab - ba$. It is readily verified (by counting like terms) that

$$\psi_{2n}(a_1, \ldots, a_{2n}) = 2^n F_{2n}(a_1, \ldots, a_{2n}) . \tag{20.5}$$

We remind the reader that the coefficients of the powers λ^{n-k} in the characteristic polynomial $\Sigma(-1)^k \alpha_k \lambda^{n-k}$ of a complex square matrix a are given by Newton's formula

$$\alpha_k = \frac{1}{k} \sum_{j=1}^{k} (-1)^{j-1} \alpha_{k-j} tr(a^j)$$

(see [19]). It follows that the coefficients α_k have the form

$$\alpha_k = \sum_p q_p \ tr(a^{p_1}) \ldots tr(a^{p_j}) , \tag{20.6}$$

where $q_p \in \mathbb{C}$ and $p = (p_1, \ldots, p_j)$ with $p_1 \le \ldots \le p_j$.

We shall use the following notations. If $\Phi : \mathcal{K}^R \to \mathbb{C}$, where $R \in \mathbb{N}$, is a map, then we let $\Delta\Phi$ denote the map $\mathcal{K}^{R+1} \to \mathbb{C}$ defined by the formula

$$\Delta\Phi(c_1, \ldots, c_{R+1}) = \Phi(c_1 + c_{R+1}, c_2, \ldots, c_R) - \Phi(c_{R+1}, c_2, \ldots, c_R) - \Phi(c_1, c_2, \ldots, c_R).$$

Consider the map $\Phi_0 : \mathcal{K} \to \mathbb{C}$,

$$\Phi_0(x) = x^n + \sum_{k=1}^{n} \sum_p q_p tr(x^{p_1}) \ldots tr(x^{p_j}), \tag{20.7}$$

where the sum $\sum_p$ runs over the same multiindices $p = (p_1, \ldots, p_j)$ as in formula (20.6), and the allied map $\Delta^{n-1}\Phi_0 : \mathcal{K}^n \to \mathbb{C}$. Since the trace possesses the additivity property,

$$0 = \Delta^{n-1}\Phi_0(x_1, \ldots, x_n) = \sum_\sigma x_{\sigma(1)} x_{\sigma(2)} \cdots x_{\sigma(n)}$$

$$+ \sum_{k=1}^{n} \sum_p \sum_\sigma q_p tr(x_{\sigma(1)} \cdots x_{\sigma(p_1)}) \ldots x_{\sigma(n-k+1)} x_{\sigma(n-k+2)} \cdots x_{\sigma(n)}. \tag{20.8}$$

Upon replacing every element x_i by $[a_{2i-1}, a_{2i}]$ and using formula (20.3) we get $0 = 2^n F_{2n}(a_1, \ldots, a_{2n}) + \rho$, where ρ designates the sum of the terms

$$q_p tr\big(F_{2p_1}(a_{2\sigma(1)-1}, a_{2\sigma(1)}), \ldots, a_{2\sigma(p_1)-1}, a_{2\sigma(p_1)}\big) \cdot \ldots \cdot F_{2(n-k)}\big(a_{2\sigma(k+1)-1}, \ldots, a_{2\sigma(n)}\big) \tag{20.9}$$

corresponding to distinct $\sigma \in S_n$. Notice that the matrix $z = \sum_\sigma [a_{2\sigma(1)-1}, a_{2\sigma(1)}] \cdots$ $[a_{2\sigma(n)-1}, a_{2\sigma(n)}]$ combines only some of the terms appearing in the sum $\Psi_{2n}(a_1, \ldots, a_{2n})$. The matrix $\Psi_{2n}(a_1, \ldots, a_{2n}) + \rho$ is a linear combination of the terms obtained from z by permutations of the indices (the details are left to the reader). It remains to observe that

$$
2\,tr F_{2k}(a_1, \ldots, a_{2k}) = tr\left(\sum_{i=1}^{2k}(-1)^{i-1}a_i F_{2k-1}(a_1, \ldots, a_{i-1}, a_{i+1}, \ldots, a_{2k})\right)
$$
$$
+ tr\left(\sum_{i=1}^{2k}(-1)^{2k-1} F_{2k-1}(a_1, \ldots, a_{i-1}, a_{i+1}, \ldots, a_{2k})a_i\right)
$$
$$
= \sum_{i=1}^{2k}(-1)^{i-1}tr[a_i, F_{2k-1}(a_1, \ldots, a_{i-1}, a_{i+1}, \ldots, a_{2k})] = 0.
$$

This implies, using (20.9), that $\rho(a_1, \ldots, a_{2n}) = 0$, and hence that $F_{2n}(a_1, \ldots, a_{2n}) = 0$ for any collection of matrices $a_1, \ldots, a_{2n} \in \mathbb{C}^{n\times n}$.

$\square$

We remark that F_{2n} is a polynomial of degree $2n$. Polynomials of degree less than $2n$ cannot vanish identically on $\mathbb{C}^{n\times n}$.

THEOREM 20.2. *The algebra $\mathbb{C}^{n\times n}$ satisfies no polynomial identities of degree less than $2n$.*

For the proof we need

LEMMA 20.1. *If the algebra $\mathcal{K}$ satisfies a polynomial identity f of degree d, then it satisfies a multilinear identity of degree $\leq d$.*

PROOF. If $f(x_1, \ldots, x_k) = 0$ for every collection $x_1, \ldots, x_k \in \mathcal{K}$ then $g(x_1, \ldots, x_k, x_{k+1}) = 0$, where

$$
g(x_1, x_2, \ldots, x_{k+1}) = f(x_1 + x_{k+1}, x_2, \ldots, x_k) - f(x_1, \ldots, x_k) - f(x_{k+1}, x_2, \ldots, x_k).
$$

Every such step diminishes the degree of the corresponding variable and does not increase the overall degree of the polynomial. After a finite number of steps we arrive at a multilinear identity.

$\square$

PROOF OF THEOREM 20.2. If $\mathbb{C}^{n \times n}$ satisfies a polynomial identity of degree d, then it satisfies a multilinear identity of degree $m \leq d$:

$$x_1 x_2 \ldots x_m + \sum \alpha_\sigma x_{\sigma(1)} \ldots x_{\sigma(m)}, \qquad (20.10)$$

in which the sum is taken over all nonidentical permutations. Now set $x_{2p-1} = e_{p,p}$ and $x_{2p} = e_{p,p+1}$, where e_{pq} are the matrix units, i.e., the matrices with the entries $(e_{pq})_{jk} = \delta_{jp}\delta_{kq}$ (this is possible because $m < 2n$). It is readily checked that $x_{\sigma(1)} \ldots x_{\sigma(m)} = 0$ for every nonidentical permutation σ and all $x_1, \ldots, x_m \neq 0$. But this contradicts identity (20.10).

$\square$

21. A MATRIX ANALOGUE OF GELFAND'S TRANSFORMATION

Let $\mathcal{K}$ be a Banach algebra with identity over $\mathbb{C}$, $\mathcal{R}(\mathcal{K})$ the radical of $\mathcal{K}$ (see Sec.13), $G\mathcal{K}$ the group of invertible elements of $\mathcal{K}$, and $\mathcal{M}$ the set of all two-sided maximal ideals of $\mathcal{K}$. We may assume, without loss of generality, that

$$\|e\| = 1 \quad \text{and} \quad \|xy\| \leq \|x\|\|y\| \qquad (21.1)$$

for all $x, y \in \mathcal{K}$.

We shall write $\mathcal{K} \in \mathcal{F}_m$ if the standard identity $F_m(a_1, \ldots, a_m) = 0$ is fulfilled for every collection $a_1, \ldots, a_m$ of elements of $\mathcal{K}$. By Theorem 20.1, $\mathbb{C}^{n \times n} \in \mathcal{F}_{2n}$.

THEOREM 21.1. *Suppose that $\mathcal{K} \in \mathcal{F}_{2n}$. Then:*

a) *for every ideal $M \in \mathcal{M}$ the quotient algebra $\mathcal{K}_M = \mathcal{K}/M$ is isomorphic to $\mathbb{C}^{\ell \times \ell}$, where $\ell = \ell(M) \leq n$;*

b) *if $\nu_M = \xi_M \eta_M$, where η_M and ξ_M designate the natural homomorphism $\mathcal{K} \to \mathcal{K}_M$ and the isomorphism $\mathcal{K}_M \to \mathbb{C}^{\ell \times \ell}$, respectively, then*

$$x \in G\mathcal{K} \iff (\det \nu_M(x) \neq 0 \quad \forall\, M \in \mathcal{M}); \qquad (21.2)$$

c) *the radical* $\mathcal{R}(\mathcal{K})$ *coincides with the intersection of all two-sided maximal*
ideals of $\mathcal{K}$: $\mathcal{R}(\mathcal{K}) = \bigcap_{M \in \mathcal{M}} M$.

For $n = 1$ the condition $\mathcal{K} \in \mathcal{F}_2$ means that $\mathcal{K}$ is commutative and the assertions of Theorem 21.1 become well-known results from the theory of commutative Banach
algebras [20, Theorems 3, 3', and 5]. To prove the theorem we need a number of auxiliary
propositions.

First we remind the reader of the notion of a left regular representation of
algebra $\mathcal{K}$ [62]. Let $\mathcal{I}$ be a left maximal ideal of $\mathcal{K}$, $E = \mathcal{K}/\mathcal{I}$, and $\tau: \mathcal{K} \to E$ the natural
map. To each element $a \in \mathcal{K}$ we assign the bounded linear operator $R_a: E \to E$ defined
by the rule $R_a(\tau(x)) = \tau(ax)$. The homomorphism $\nu: \mathcal{K} \to \mathcal{L}(E)$ which sends a into R_a
is called a *left regular representation* of the algebra $\mathcal{K}$. It is readily verified that $\|\nu\| = 1$
and that ν is an algebraically irreducible representation of $\mathcal{K}$ (i.e., if $E_0 \neq \{0\}$ is a linear
manifold in E invariant under all operators R_a, then $E_0 = E$). In this section we shall
write $\hat{x}$ instead of $\tau(x)$.

We give one of the variants of Schur's lemma.

LEMMA 21.1. *Let B be a linear operator in E (which is not assumed a priori
to be bounded). If* $BR_a = R_a B$ *for all* $a \in \mathcal{K}$, *then B is a scalar operator:* $B = \lambda I$.

PROOF. First we show that B is bounded. Let $\hat{x} \in E$ and let $a \in \hat{x}$ be
an element such that $\|a\| \leq 2\|\hat{x}\|$. Then $\|B\hat{x}\| = \|BR_a\hat{e}\| = \|R_a B\hat{e}\| \leq \|a\|\|B\hat{e}\| = 2\|B\hat{e}\| \cdot \|\hat{x}\|$. Now let us show that if $B \neq 0$ then $B \in GL(E)$. Since $R_a(Im\ B) \subset Im\ B$
and $Im\ B \neq \{0\}$, $Im\ B = E$. Moreover, $R_a(Ker\ B) \subset Ker\ B$ and $Ker\ B \neq E$, and
hence $Ker\ B = \{0\}$, i.e., $B \in GL(E)$.

Let λ be a point of the spectrum of B. Since $B - \lambda I \notin GL(E)$, the above
argument shows that $B - \lambda I = 0$.

$\square$

In the next two lemmas we prove a result which is known in algebra as the
density theorem [27,67].

LEMMA 21.2. *Let E_0 be a finite-dimensional linear manifold in E and let*
$\hat{x} \notin E_0$. *Then there is an $a \in \mathcal{K}$ such that $R_a(E_0) = \{0\}$ and $R_a \hat{x} \neq 0$.*

PROOF. We proceed by induction on the dimension of E_0. The assertion of the lemma is obviously true when $dim\ E_0 = 0$. Suppose that it is true for $dim\ E_0 = k$. Set $\tilde{E} = E_0 + \mathbb{C}\hat{\omega}$, where $\hat{\omega} \notin E_0$, and let $\mathcal{X} = \{R_a : R_a(E_0) = \{0\}\}$. It is easily seen that the set $\{R_a\hat{\omega} : R_a \in \mathcal{X}\}$ is linear, invariant under all operators R_b with $b \in \mathcal{K}$, and does not reduce to zero. It follows from the algebraic irreducibility of representation ν that $\{R_a\hat{\omega} : R_a \in \mathcal{X}\} = E$. Now suppose that the lemma is not valid for $\tilde{E}$, i.e., there is a $\hat{z} \notin \tilde{E}$ such that $R_a\hat{z} = 0$ for every a satisfying $R_a(\tilde{E}) = 0$. Consider the operator B which acts in E according to the rule $B\hat{x} = R_a\hat{z}$, where $R_a\hat{\omega} = \hat{x}\ (R_a \in \mathcal{X})$. It is readily checked that B is correctly defined and satisfies the conditions of Lemma 21.1. Consequently, $B = \lambda I$. We thus have, for every $R_a \in \mathcal{X}$, $R_a\hat{z} = BR_a\hat{\omega} = \lambda R_a\hat{\omega}$, i.e., $R_a(\hat{z} - \lambda\hat{\omega}) = 0$. By the inductive assumption, if $R_a\hat{\xi} = 0$ for all $R_a \in \mathcal{X}$, then $\hat{\xi} \in E_0$. Therefore, $\hat{z} - \lambda\hat{\omega} \in E_0$, which contradicts our choice $\hat{z} \notin \tilde{E}$.

$\square$

LEMMA 21.3. *Let* $\{\hat{v}_k\}_1^n \subset E$, $\{\hat{u}_k\}_1^n \in E$, *and suppose that* $\hat{u}_1, \dots, \hat{u}_n$ *are linearly independent. Then there exists an element* $a \in \mathcal{K}$ *such that* $R_a\hat{u}_k = \hat{v}_k$, $k = 1, \dots, n$.

PROOF. It follows from Lemma 21.2 that for each $k = 1, \dots, n$ there is an $a_k \in \mathcal{K}$ such that $R_{a_k}\hat{u}_k \neq 0$ and $R_{a_k}\hat{u}_m = 0$ for $m \neq k$. Consider the linear manifold $E_k = \{R_x R_{a_k}\hat{u}_k : x \in \mathcal{K}\}$. Since $R_a(E_k) \subset E_k$ for every $k = 1, \dots, n$, and $R_e R_{a_k}\hat{u}_k \neq 0$, we have $E_k = E$, $k = 1, \dots, n$, and hence there is an $x_k \in \mathcal{K}$ such that $R_{x_k a_k}\hat{u}_k = \hat{v}_k$ and $R_{x_k a_k}\hat{u}_m = 0\ (m \neq k)$. The element $a = \sum_{k=1}^n x_k a_k$ has the desired property.

$\square$

LEMMA 21.4 (Kaplansky). *If* $\mathcal{K} \in \mathcal{F}_{2n}$ *then*

$$\dim E \leq n \quad and \quad \{R_a\}_{a \in \mathcal{K}} = \mathcal{L}(E). \tag{21.3}$$

PROOF. Suppose that $\dim E > n$, and let $\hat{u}_1, \dots, \hat{u}_{n+1}$ be linearly independent elements of E. By Lemma 21.3, there exist $a_k, b_k \in \mathcal{K}$ such that

$$R_{a_k}\hat{u}_i = \delta_{ik}\hat{u}_i \qquad (k, i = 1, 2, \dots, n)$$

and

$$R_{b_k}\hat{u}_i = \delta_{ik}\hat{u}_{i-1} \qquad (k, i = 1, 2, \dots, n+1).$$

Set $c_{2m} = b_{m+1}$ and $c_{2m-1} = a_m$ $(m = 1, \ldots, n)$. It is readily checked that $R_{c_1} R_{c_2} \cdots R_{c_{2n}} \hat{u}_{n+1} = \hat{u}_1 \neq 0$ and that $R_{c_{\sigma(1)}} R_{c_{\sigma(2)}} \cdots R_{c_{\sigma(2n)}} \hat{u}_{n+1} = 0$ for every nonidentical permutation σ of the numbers $1, 2, \ldots, 2n$. Consequently, $F_{2m}(R_{c_1}, \ldots, R_{c_{2n}}) \neq 0$, which contradicts the assumption that $\mathcal{K} \in \mathcal{F}_{2n}$.

□

Algebra $\mathcal{K}$ is called *primitive* [62] if it contains a left maximal ideal $\mathcal{I}$ for which the corresponding left regular representation $\nu \colon \mathcal{K} \to \mathcal{L}(\mathcal{K}/\mathcal{I})$ is an isomorphism.

LEMMA 21.5. *Algebra $\mathcal{K}$ is primitive if and only if there is a left maximal ideal in $\mathcal{K}$ which contains no nonnull two-sided ideals of $\mathcal{K}$.*

PROOF. Suppose $\mathcal{K}$ is primitive and let $\mathcal{I}$ be a left maximal ideal for which the left regular representation ν is an isomorphism. Let $\mathcal{J} \subset \mathcal{I}$ be a two-sided ideal of $\mathcal{K}$ and $a \in \mathcal{J}$. Then $R_a \hat{x} = \widehat{ax} = 0$ for all x $(ax \in \mathcal{J} \subset \mathcal{I})$. Since ν is an isomorphism, $a = 0$.

Conversely, suppose that $\mathcal{I}$ contains no nonnull two-sided ideals of $\mathcal{K}$. If $\nu(a) = 0$, then $ax \in \mathcal{I}$ for all $x \in \mathcal{K}$. Set $\mathcal{J} = \{\sum_k y_k a x_k : x_k, y_k \in \mathcal{K}\}$. Then it is readily checked that $\mathcal{J}$ is a two-sided ideal of $\mathcal{K}$ and $\mathcal{J} \subset \mathcal{I}$. Consequently, $\mathcal{J} = \{0\}$, and hence $a = 0$.

□

LEMMA 21.6. *Every primitive Banach algebra $\mathcal{K}$ over $\mathbb{C}$ which satisfies a standard identity (i.e., $\mathcal{K} \in \mathcal{F}_{2n}$) is isomorphic to $\mathbb{C}^{\ell \times \ell}$, where $\ell \leq n$.*

PROOF. Since $\mathcal{K}$ is primitive, $\mathcal{K}$ is isomorphic to the algebra $\{R_a : a \in \mathcal{K}\}$. By Lemma 21.4, $\mathcal{K}$ is isomorphic to $\mathcal{L}(E)$ and $\dim E \leq n$.

□

We shall also need

LEMMA 21.7. *If $\mathcal{K} \in \mathcal{F}_{2n}$, then every left maximal ideal $\mathcal{I}$ of $\mathcal{K}$ contains a two-sided maximal ideal of $\mathcal{K}$.*

PROOF. Let ν denote the left regular representation corresponding to $\mathcal{I}$ and $\mathcal{J} = \mathrm{Ker}\, \nu$. Obviously, $\mathcal{J}$ is a closed two-sided ideal of $\mathcal{K}$ and $\mathcal{J} \subset \mathcal{L}$. The quotient algebra $\mathcal{K}_{\mathcal{J}} = \mathcal{K}/\mathcal{J}$ is primitive. In fact, if $\mathcal{T}$ is a two-sided ideal of $\mathcal{K}$ and $\mathcal{T} \subset \mathcal{I}$, then $R_s \hat{x} = \widehat{sx} = 0$ for all $s \in \mathcal{T}$ and hence $R_s = 0$, i.e., $s \in \mathrm{Ker}\, \nu = \mathcal{J}$. Therefore, $\mathcal{J}$ is the largest two-sided ideal of $\mathcal{K}$ contained in $\mathcal{I}$. This shows that in the quotient algebra $\mathcal{K}_{\mathcal{J}}$ the left maximal ideal $\mathcal{I}/\mathcal{J}$ contains no two-sided ideals. Since $\mathcal{K}_{\mathcal{J}} \in \mathcal{F}_{2m}$, $\mathcal{K}_{\mathcal{J}}$ is isomorphic to $\mathbb{C}^{\ell \times \ell}$ by Lemma 21.6, and hence is a simple algebra. Consequently, $\mathcal{J}$ is a two-sided maximal ideal of the algebra $\mathcal{K}$.

□

PROOF OF THEOREM 21.1. 1) For every maximal ideal M the quotient-algebra $\mathcal{K}/M$ is simple and belongs to $\mathcal{F}_{2n}$. By Lemma 21.6, $\mathcal{K}/M \cong \mathbb{C}^{\ell \times \ell}$, with $\ell \leq n$.

2) The implication $\Longrightarrow$ in (21.2) is obvious. Let us prove $\Longleftarrow$. Suppose that $det\nu_M(x) \neq 0$ for all $M \in \mathcal{M}$. Then $x_M \in G\mathcal{K}_M$ for all $M \in \mathcal{M}$. We first show that x is left invertible in $\mathcal{K}$. Assuming the contrary, x must belong to a left maximal ideal $\mathcal{I}$. By Lemma 21.7 there is a maximal ideal $M \subset \mathcal{I}$. We let η_M denote the natural homomorphism $\mathcal{K} \to \mathcal{K}_M$, and we set $\mathcal{I}_M = \eta_M(\mathcal{I})$. Since $M \subset \mathcal{I}$, $\eta_M(x_0) \in \mathcal{I}_M$ if and only if $x_0 \in \mathcal{I}$. Consequently, $x_M \in \mathcal{I}_M$, which contradicts the invertibility of x_M in $\mathcal{K}_M$. Now let us show that x is right invertible. Since x is left invertible, there is a $u \in \mathcal{K}$ such that $ux = e$. It follows from the equality $u_M x_M = e_M$ and the invertibility of x_M that $x_M u_M = e_M$ for all $M \in \mathcal{M}$. By Lemma 21.7, $r = xu - e$ belongs to every left maximal ideal, and hence to the radical of $\mathcal{K}$. This implies the invertibility of the element $xu = e+r$ and hence the right invertibility of x.

3) By Lemma 21.7, the intersection of all maximal two-sided ideals is contained in the radical $\mathcal{R}(\mathcal{K})$. Conversely, let $x \in \mathcal{R}(\mathcal{K})$. Then $x_M \in \mathcal{R}(\mathcal{K}_M)$, but $\mathcal{K}_M$ are semisimple algebras, and hence $x_M = 0$ for all $M \in \mathcal{M}$, i.e., $x \in \bigcap_{M \in \mathcal{M}} M$.

□

22. SUFFICIENT FAMILIES OF n-DIMENSIONAL REPRESENTATIONS

Let $\mathcal{K}$ be a Banach algebra. We say that $\mathcal{K}$ possesses a *sufficient family of n-dimensional representations*, and write $\mathcal{K} \in \pi_n$, if there exists a family $\{\nu_M\}_{M \in \alpha}$ of homomorphisms $\nu_M: \mathcal{K} \to \mathbb{C}^{\ell \times \ell}$ (with $\ell = \ell(M) \leq n$), such that

$$x \in G\mathcal{K} \Longleftrightarrow (det\ \nu_M(x) \neq 0 \quad \forall\, M \in \alpha). \qquad (22.1)$$

LEMMA 22.1. *If* $\mathcal{K}/\mathcal{R}(\mathcal{K}) \in \pi_n$, *then* $\mathcal{K} \in \pi_n$.

PROOF. Let $\{\tilde{\nu}_M\}_{M \in \alpha}$ be a sufficient family of n-representations of the algebra $\tilde{\mathcal{K}} = \mathcal{K}/\mathcal{R}(\mathcal{K})$. Set $\nu_M(x) = \tilde{\nu}_M(\tilde{x})$ for $x \in \tilde{x}$. Then

$$(det\ \nu_M(x) = 0\ \ \forall\ M \in \alpha) \Longleftrightarrow \tilde{x} \in G\tilde{\mathcal{K}} \Longleftrightarrow x \in G\mathcal{K}.$$

$\square$

THEOREM 22.1. *Let $\mathcal{K}$ be a Banach algebra. Then $\mathcal{K} \in \pi_n$ if and only if $F_{2n}(a_1, \ldots, a_{2n}) \in \mathcal{R}(\mathcal{K})$ for every collection of elements $a_1, \ldots, a_{2n} \in \mathcal{K}$.*

PROOF. If $F_{2n}(a_1, \ldots, a_{2n}) \in \mathcal{R}(\mathcal{K})$ for every collection $a_1, \ldots, a_{2n} \in \mathcal{K}$, then $\mathcal{K}/\mathcal{R}(\mathcal{K}) \in \mathcal{F}_{2n}$. By Theorem 21.1, $\mathcal{K}/\mathcal{R}(\mathcal{K}) \in \pi_n$, and hence $\mathcal{K} \in \pi_n$ by Lemma 22.1. Conversely, suppose that $\mathcal{K} \in \pi_n$ and let $a_1, \ldots, a_{2n} \in \mathcal{K}$, $x = F_{2n}(a_1, \ldots, a_{2n})$, u an arbitrary element of $\mathcal{K}$, and $z = e + ux$. Then $\nu_M(z) = \nu_M(e) + F_{2n}(\nu(a_1), \ldots, \nu(a_{2n}))$ for every homomorphism ν_M from a sufficient family of n-dimensional representations of $\mathcal{K}$. Since $\mathbb{C}^{\ell \times \ell} \in \mathcal{F}_{2n}$ for all $\ell \leq n$, it follows that $det\ \nu_M(z) = 1$ for all M. Consequently, $z \in G\mathcal{K}$, and hence $x \in \mathcal{R}(\mathcal{K})$.

$\square$

We recall that $\mathcal{K}$ is called a PI-*algebra* (polynomial-identity algebra) if there exists a noncommutative polynomial $f(x_1, \ldots, x_k)$, $f \neq 0$, such that $f(a_1, \ldots, a_k) = 0$ for arbitrary $a_1, \ldots, a_k \in \mathcal{K}$ (see [27]). The polynomial f is also referred to as an *identity* on $\mathcal{K}$.

Every commutative algebra is a PI-algebra: $f(a, b) = ab - ba$ works. The matrix algebras $\mathbb{C}^{n \times n}$ are PI-algebras with the standard identity F_{2n}. On the algebra $\mathbb{C}^{2 \times 2}$ the identity $[x, [y, z]^2] = 0$ also holds (it is readily checked that $[y, z]^2$ is a scalar multiple of the identity matrix).

THEOREM 22.2. *In order for the Banach algebra $\mathcal{K}$ to belong to the class π_n for some n it is necessary and sufficient that $\tilde{\mathcal{K}} = \mathcal{K}/\mathcal{R}(\mathcal{K})$ be a PI-algebra.*

PROOF. The necessity follows from the preceding theorem. Now suppose that $\tilde{\mathcal{K}}$ is a PI-algebra. Then $\tilde{\mathcal{K}}$ satisfies a polynomial identity (see Lemma 20.1). In the proof of Theorem 21.1 we used only this property of standard polynomials. It follows from this theorem that $\tilde{\mathcal{K}} \in \pi_n$ for some value of n. Now apply Lemma 22.1.

$\square$

COROLLARY 22.1. *If $\mathcal{K}$ is a subalgebra of $\mathcal{L}(E)$ which contains all the rank-one operators, then $\mathcal{K} \notin \pi_n$ for any value of n.*

PROOF. We first show that $\mathcal{K}$ is semisimple. Assuming the contrary, let $A \in \mathcal{R}(\mathcal{K})$ be such that $Ax_0 = z_0 \neq 0$ for some $x_0 \in E$. Set $Kx = f(x)x_0$, where $f(x_0) = 1$, and $T = AK$. Since $\mathcal{R}(\mathcal{K})$ is a two-sided ideal, $T \in \mathcal{R}(\mathcal{K})$, and hence $I - T \in G\mathcal{K}$. But $(I - T)z_0 = 0$, which is a contradiction. Now suppose that $\mathcal{K} \in \pi_n$ for some n. Then $\mathcal{K} \in \mathcal{F}_{2n}$. But $\mathcal{K}$ contains all the finite-rank operators, and hence, by Theorem 20.2, it cannot satisfy polynomial identities.

$\square$

The next theorem describes the minimal order of sufficient families of representations that a given algebra admits.

THEOREM 22.3. *Suppose $\mathcal{K} \in \pi_n$. Then*

$$\min\{m\colon \mathcal{K} \in \pi_m\} = \max\{\ell(M)\colon \mathcal{K}_M \cong \mathbb{C}^{\ell(M)\times\ell(M)}\}.$$

PROOF. If $\mathcal{K} \in \pi_m$, then $\mathcal{K}/\mathcal{R}(\mathcal{K}) \in \mathcal{F}_{2m}$, and hence $\mathcal{K}_M \in \mathcal{F}_{2m}$. By Theorem 20.2, $\ell(M) \leq m$, and hence

$$\min\{m\colon \mathcal{K} \in \pi_m\} \geq \max\{\ell(M)\colon \mathcal{K}_M \cong \mathbb{C}^{\ell(M)\times\ell(M)}\}.$$

The opposite inequality follows from Theorem 20.1.

$\square$

COROLLARY 22.1. *Suppose $\mathcal{K} \in \pi_n$ and let $\{\nu_M\}_{M\in\alpha}$ be a sufficient family of representations. Then $\mathcal{K} \notin \pi_m$ for $m < n$ if and only if there is an $M \in \alpha$ such that $Im\,\nu_M = \mathbb{C}^{n\times n}$.*

$\square$

Equality (22.2) shows that the order of sufficient families of representations that an algebra $\mathcal{K}$ admits depends on the number $p = \max\{\ell(M)\colon \mathcal{K}_M \cong \mathbb{C}^{\ell(M)\times\ell(M)}\}$. The following question arises naturally: suppose that for every $M \in \mathcal{M}$ the quotient algebra $\mathcal{K}_M$ is isomorphic to $\mathbb{C}^{\ell(M)\times\ell(M)}$ and $\ell(M) \leq n$, where n is fixed; does this imply that $\mathcal{K} \in \pi_n$? The answer is negative, as the following example shows:

Example 22.1. Let $E = \ell_2$ and $\mathcal{K} = \{\lambda I + T : \lambda \in \mathbb{C}, T \in \mathcal{T}(E)\}$. Then obviously $\mathcal{T}(E)$ is a two-sided maximal ideal in $\mathcal{K}$. We show that $\mathcal{K}$ contains no other closed two-sided ideals. Let $\mathcal{J}$ be a closed two-sided ideal of $\mathcal{K}$ containing an operator $A \neq 0$, and let K be an arbitrary rank-one operator: $Kx = f(x)z$. Then there is an x_0 such that $y = Ax_0 \neq 0$. Consider the two rank-one operators $Bx = f(x)x_0$ and $Cx = g(x)z$, where $g(y) = 1$. Then $CAB = K \in \mathcal{J}$. We have thus shown that $\mathcal{J}$ contains all rank-one operators, and hence $\mathcal{J} \supset \mathcal{T}(E)$. Since $\mathcal{T}(E)$ is a maximal ideal, it follows that $\mathcal{J} = \mathcal{T}(E)$. Therefore, $\mathcal{K}$ contains the unique two-sided maximal ideal $\mathcal{T}(E)$ and the quotient algebra $\mathcal{K}/\mathcal{T}(E)$ is isomorphic to $\mathbb{C}$. But, by Corollary 22.1, $\mathcal{K} \notin \pi_n$ for any n.

Example 22.2. Let $E = \ell_p$ $(p \geq 1)$, $m \in \mathbb{N}$, and let $\mathcal{K}$ be the subalgebra of $\mathcal{L}(E)$ generated by all operators B, $B\{\xi_n\}_1^\infty = \{\lambda_n \xi_n\}_1^\infty$, of multiplication by convergent sequences of complex numbers λ_n, and the Carleman shift operator W, $W\{\xi_n\}_1^\infty = \{\eta_n\}_1^\infty$, where $\eta_n = \xi_{n-1}$ if $n \not\equiv 1 \pmod{m}$ and $\eta_{km+1} = \xi_{m(k+1)}$, $k = 0, 1, 2, \ldots$.

The matrix of each operator $A \in \mathcal{K}$ in the standard basis of ℓ_p has the block-diagonal form: $A = \mathrm{diag}(C_1, C_2, \ldots)$, where $C_j \in \mathbb{C}^{m \times m}$. Since $\mathbb{C}^{m \times m} \in \mathcal{F}_{2m}$, it follows that $\mathcal{K} \in \mathcal{F}_{2m}$. By Theorem 22.1, $\mathcal{K} \in \pi_m$.

It is interesting to note that the matrices C_j that were used to establish that $\mathcal{K} \in \pi_m$ are, to a certain extent, but not completely, responsible for the invertibility of the operator A. Specifically, the conditions $det\ C_j \neq 0$ are necessary for the invertibility of A (which is obvious), but not sufficient (as the example of the operator $A\{\xi_n\}_1^\infty = \{\frac{1}{n}\xi_n\}_1^\infty$ shows). This implies that the homomorphisms $\nu_j(A) = C_j$ do not form a sufficient family. Later we shall shall return to Example 22.2 after we present some methods for constructing sufficient families.

23. A METHOD FOR CONSTRUCTING SUFFICIENT FAMILIES OF REPRESENTATIONS

In the preceding sections we found a condition for the existence of a sufficient family of n-representations. Here we construct such a family for a certain class of Banach algebras.

Suppose that in algebra $\mathcal{K}$ there exist elements $p_1, \ldots, p_n$, and v such that

$$
\left.
\begin{aligned}
p_j p_k &= \delta_{jk} p_j, \qquad \textstyle\sum_{j=1}^n p_j = e, \\
&\text{and} \\
v^{-1} p_j v &= p_{j+1} \qquad (p_{n+1} = p_1)
\end{aligned}
\right\}
\tag{23.1}
$$

(where δ_{jk} is the Kronecker symbol).

We let $\mathcal{K}_0$ denote the subalgebra of $\mathcal{K}$ consisting of the elements of the form $p_1 x_1 p_1 + p_2 x_2 p_2 + \ldots + p_n x_n p_n$:

$$
\mathcal{K}_0 = \sum_{j=1}^n p_j \mathcal{K} p_j.
\tag{23.2}
$$

THEOREM 23.1. $\mathcal{K}$ is isomorphic to a Banach subalgebra of $\mathcal{K}_0^{n \times n}$.

PROOF. For each $x \in \mathcal{K}$ we put

$$
\mu(x) = \left[\sum_{j=1}^n p_j v^{m-1} x v^{1-k} p_j \right]_{m,k=1}^n \in \mathcal{K}_0^{n \times n}.
\tag{23.3}
$$

It is readily checked that $\mu(x+y) = \mu(x) + \mu(y)$ and $\|\mu(x)\| \le c_1 \|x\|$. Moreover,

$$
\mu(x)\mu(y) = \left[\sum_{j=1}^n p_j v^{m-1} x v^{1-k} p_j \right] \left[\sum_{j=1}^n p_j v^{m-1} y v^{1-k} p_j \right]
$$

$$
= \left[\sum_{j=1}^n \sum_{q=1}^n p_j v^{m-1} x v^{1-q} p_j v^{q-1} y v^{1-k} p_j \right]
$$

$$
= \left[\sum_{j=1}^n \sum_{q=1}^n p_j v^{m-1} x p_{j+q-1} y v^{1-k} p_j \right]
$$

$$
= \left[\sum_{j=1}^n p_j v^{m-1} x y v^{1-k} p_j \right] = \mu(xy)
$$

(for $s > n$ we put $p_s = p_k$, where $1 \le k \le n$, $s \equiv k \pmod n$). We next show that $\|x\| \le c_2 \|\mu(x)\|$. Let $u_k = \sum p_j x v^{1-k} p_j$ be the entries on the first row of the matrix (23.3). Then

$$
\sum_k u_k v^{k-1} = \sum_{j,k} p_j x v^{1-k} p_j v^{k-1} = \sum_{j,k} p_j x p_{j+k-1} = x.
$$

Hence, $\|x\| \leq c \max \|u_k\| \leq c_2 \|\mu(x)\|$. This implies, in particular, that the image of μ is closed, as claimed.

$\square$

THEOREM 23.2. *If the assumptions of the previous theorem are in force, then*

$$x \in G\mathcal{K} \Longleftrightarrow \mu(x) \in G\mathcal{K}_0^{n \times n}. \qquad (23.4)$$

PROOF. Let $\mu(x) \in G\mathcal{K}_0^{n \times n}$; then $\mu(x) \in G\mathcal{K}^{n \times n}$. Consider the elements $a, b \in \mathcal{K}^{n \times n}$ defined as

$$a = [\varepsilon^{(m-1)(k-1)} v^{1-k}]_{m,k=1}^n$$

and

$$b = [\varepsilon^{(1-m)(k-1)} v^{m-1}]_{m,k=1}^n,$$

where $\varepsilon = \exp(2\pi i/n)$. Then

$$a\mu(x)b = \left[\sum_{q,j,s=1}^n \varepsilon^{(m-1)(q-1)} v^{1-q} p_j v^{q-1} x v^{1-s} p_j \varepsilon^{(1-s)(k-1)} v^{s-1} \right]_{m,k=1}^n$$

$$= \left[\sum_{q,j,s=1}^n \varepsilon^{(m-1)(q-1)+(1-s)(k-1)} p_{q+j-1} x p_{j+s-1} \right]_{m,k=1}^n$$

$$= \left[\sum_{j=1}^n \varepsilon^{j(k-1)-j(m-1)} \left(\sum_{r=1}^n \varepsilon^{r(m-1)} p_r \right) x \left(\sum_{r=1}^n \varepsilon^{r(1-k)} p_r \right) \right]_{m,k=1}^n.$$

Consequently,

$$a\mu(x)b = n \ diag(x_1, \ldots, x_n), \qquad (23.5)$$

where

$$x_m = \left(\sum_{k=1}^n \varepsilon^{k(m-1)} p_k \right) x \left(\sum_{k=1}^n \varepsilon^{k(1-m)} p_k \right).$$

In particular, $x_1 = x$. Since $ab = ba = n[\delta_{mk} e]$, the matrices a and b belong to $G\mathcal{K}^{n \times n}$, and, by (23.5), $x \in G\mathcal{K}$. The implication $\Longrightarrow$ in (23.4) is obvious.

$\square$

We define a class ω_n of Banach algebras for which we shall construct below a sufficient family of n-dimensional representations. We say that $\mathcal{K} \in \omega_n$ $(n \geq 1)$ if there exist $n + 1$ elements $p_1, \ldots, p_n, v$ in $\mathcal{K}$ which satisfy relations (23.1) and such that the subalgebra $\mathcal{K}_0 = \sum_{j=1}^{n} p_j \mathcal{K} p_j$ is commutative.

For $n = 1$, $\mathcal{K} \in \omega_1$ if and only if $\mathcal{K}$ is commutative (to see this take $p_1 = e$ and an arbitrary $v \in G\mathcal{K}$).

THEOREM 23.3.   *If* $\mathcal{K} \in \omega_n$, *then* $\mathcal{K} \in \pi_n$.

PROOF.   Let $x \in \mathcal{K}$, $\mu(x) = [x_{jk}]_{j,k=1}^n$, $\mathcal{M}$ the space of maximal ideals of the commutative Banach algebra $\mathcal{K}_0$, $M \in \mathcal{M}$, and $x_{jk}(M)$ the Gelfand transformation corresponding to the ideal M and the element x_{jk}. We set

$$\nu_M(x) = [x_{jk}(M)]_{j,k=1}^n \qquad (M \in \mathcal{M}). \tag{23.6}$$

According to a theorem of Gelfand [20], $(det \, \nu_M(x) \neq 0 \;\; \forall \, M \in \mathcal{M}) \Longleftrightarrow (det[x_{jk}] \in G\mathcal{K}_0)$. Since $\mathcal{K}_0$ is commutative, $det[x_{jk}] \in G\mathcal{K}_0 \Longleftrightarrow [x_{jk}] \in G\mathcal{K}_0^{n \times n}$. Using also Theorem 23.2, we see that $x \in G\mathcal{K} \Longleftrightarrow (det \, \nu_M(x) \neq 0 \;\; \forall \, M \in \mathcal{M})$. Consequently $\{\nu_M\}_{M \in \mathcal{M}}$ is a sufficient family of n-dimensional representations of the algebra $\mathcal{K}$.

$\square$

Thus, for an algebra $\mathcal{K} \in \omega_n$, the task of finding a sufficient family $\{\nu_M\}$ reduces to that of describing the set of maximal ideals of the commutative algebra $\mathcal{K}_0$.

We remark that the map (23.3) and its diagonalization (23.5) are reminiscent in form to equality (19.2) and other maps which connect pseudodifferential operators with a shift and those without shift, but having matrix coefficients (see [53]). However, there exists an essential difference: in diagonalization (23.5) the elements x_m, as well as x, are all simultaneously either invertible or not in the algebra $\mathcal{K}$. In the aforementioned works, the proof of the simultaneous invertibility of the elements x_m arising through diagonalization uses as a supplementary ingredient the specific nature of these elements. Moreover, in Sec.19 we constructed Example 19.1, in which one of the elements z_m is invertible, and the second not.

Let us examine in some examples the sufficient families of representations to which the construction indicated above leads.

Let $\Gamma_0 = \{\zeta \in \mathbb{C} : |\zeta| = 1\}$, $E = L_2(\Gamma_0)$, $(W\varphi)(t) = t^{-1}\varphi(t^{-1})$, $(H\varphi)(t) = \varphi(-t)$, and S the operator of singular integration along Γ_0. Consider the algebra

$\mathcal{K}_1 = \{\alpha_1 I + \alpha_2 S + \alpha_3 W + \alpha_4 SW : \alpha_j \in \mathbb{C}\} \subset \mathcal{L}(E)$. Setting $p_1 = \frac{1}{2}(I+S)$, $p_2 = I - p_1$, and $v = W$, it is readily established that $\mathcal{K}_1 \in \omega_2$. In this example $\mathcal{K}_0 = p_1 \mathcal{K}_1 p_1 + p_2 \mathcal{K}_1 p_2 = \{\beta_1 I + \beta_2 S : \beta_1, \beta_2 \in \mathbb{C}\}$ is commutative. If $A = \alpha_1 I + \alpha_2 S + \alpha_3 W + \alpha_4 SW$, then

$$\mu(A) = \begin{bmatrix} \alpha_1 I + \alpha_2 S & \alpha_3 I + \alpha_4 S \\ \alpha_3 I - \alpha_4 S & \alpha_1 I - \alpha_2 S \end{bmatrix}.$$

The operator $\mu(A)$ coincides with the "corresponding" operator (see Sec.19). In algebra $\mathcal{K}_0$ there are exactly two maximal ideals: $M_1 = \{\lambda p_1 : \lambda \in \mathbb{C}\}$ and $M_2 = \{\lambda p_2 : \lambda \in \mathbb{C}\}$. They generate a sufficient family of two-dimensional representations consisting of the two homomorphisms

$$\nu_1(A) = \begin{bmatrix} \alpha_1 + \alpha_2 & \alpha_3 + \alpha_4 \\ \alpha_3 - \alpha_4 & \alpha_1 - \alpha_2 \end{bmatrix}$$

and

$$\nu_2(A) = \begin{bmatrix} \alpha_1 - \alpha_2 & \alpha_3 - \alpha_4 \\ \alpha_3 + \alpha_4 & \alpha_1 + \alpha_2 \end{bmatrix}.$$

The criterion for the invertibility of the operator A is $\alpha_1^2 - \alpha_2^2 - \alpha_3^2 + \alpha_4^2 \neq 0$.

Let $\mathcal{K}_2 \subset \mathcal{L}(E)$ denote the algebra of operators of the form $AI + bH$, where $a, b \in C(\Gamma_0)$. Algebra $\mathcal{K}_2$, too, belongs to ω_2. In $\mathcal{K}_2$ one can take $p_1 = \frac{1}{2}(I + H)$, $p_2 = \frac{1}{2}(I - H)$, and $(v\varphi)(t) = t\varphi(t)$. Equalities (23.1) are easily checked. Subalgebra $\mathcal{K}_0$ consists of the operators of the form $a_0 I + b_0 H$ with even coefficients a_0, b_0, and hence is commutative. If $A = aI + bH$, then

$$\mu(A) = \begin{bmatrix} \tilde{a} + \tilde{b}H & \widetilde{t^{-1}a} - \widetilde{t^{-1}bH} \\ \widetilde{at} + \widetilde{btH} & \tilde{a} - \tilde{b}H \end{bmatrix},$$

where for a function $c(t)$ we put $\tilde{c}(t) = \frac{1}{2}(c(t) + c(-t))$.

The set $\mathcal{M}$ of maximal ideals of $\mathcal{K}_0$ may be identified with two copies of the circle Γ_0: $\mathcal{M} \approx \Gamma_0 \times \{-1, 1\}$. A sufficient family of two-dimensional homomorphisms is defined by the formulas

$$\nu_\tau(A) = \frac{1}{2} \begin{bmatrix} a(\tau) + b(\tau) + a(-\tau) + b(-\tau) & \tau^{-1}(a(\tau) - b(\tau) - a(-\tau) + b(-\tau)) \\ \tau(a(\tau) + b(\tau) - a(-\tau) - b(-\tau)) & a(\tau) - b(\tau) + a(-\tau) - b(-\tau) \end{bmatrix}$$

and

$$\nu'_\tau(A) = \frac{1}{2} \begin{bmatrix} a(\tau) - b(\tau) + a(-\tau) - b(-\tau) & \tau^{-1}(a(\tau) + b(\tau) - a(-\tau) - b(-\tau)) \\ \tau(a(\tau) - b(\tau) - a(-\tau) + b(-\tau)) & a(\tau) + b(\tau) + a(-\tau) + b(-\tau) \end{bmatrix}$$

Notice that

$$det\ \nu_\tau(A) = det\ \nu'_\tau(A) = a(\tau)a(-\tau) - b(\tau)b(-\tau).$$

The condition for the invertibility of the operator A is: $a(t)a(-t) - b(t)b(-t) \neq 0$ for all $t \in \Gamma_0$.

Let $\mathcal{K}_3$ denote the algebra generated in $\mathcal{L}(\ell_p)$ by the operators $B\{\xi_n\} = \{\lambda_n \xi_n\}$ and $W\{\xi_n\} = \{\eta_n\}$ which has been defined in Example 22.2. We recall that it is assumed that the limit $\lambda_\infty = \lim_{n \to \infty} \lambda_n$ exists. We show that $\mathcal{K}_3 \in \omega_m$. We take for p_j ($j = 1, \ldots, m$) the projectors $p_j\{\xi_n\} = \{\lambda_{nj} \xi_n\}$, where $\lambda_{nj} = 1$ if $n \equiv j \pmod m$ and $\lambda_{nj} = 0$ if $n \not\equiv j \pmod m$. The role of v is played by the operator W^{-1}. It is easily verified that relations (23.1) hold. The subalgebra $\mathcal{K}_0 = \sum p_j \mathcal{K}_3 p_j$ consists of the multiplication operators (of the form B), and hence is commutative. Thus, $\mathcal{K}_3 \in \omega_m$. The list of the maximal ideals of $\mathcal{K}_0$ is: $M_k = \{B: \lambda_k = 0\}$, $k = 1, 2, \ldots$, and $M_\infty = \{B: \lambda_\infty = 0\}$. Setting for each $A \in \mathcal{K}: A_{sk} = \sum_{j=1}^m p_j W^{1-s} A W^{k-1} p_j$, we obtain the invertibility condition:

$$A \in G\mathcal{K}_3 \Longleftrightarrow (det[A_{sk}(M)] \neq 0 \quad \forall\ M \in \{M_\infty, M_1, M_2, \ldots\}).$$

In particular, for the operator $A = B_1 + B_2 W$ (where $B_1\{\xi_n\} = \{\alpha_n \xi_n\}$, $B_2\{\xi_n\} = \{\beta_n \xi_n\}$) we have

$$\nu_{M_k}(A) = \begin{bmatrix} \alpha_{1+mk} & 0 & \cdot & \cdot & \cdot & 0 & \beta_{1+mk} \\ \beta_{2+mk} & \alpha_{2+mk} & \cdot & \cdot & & 0 & 0 \\ \cdot & \cdot & \cdot & \cdot & \cdot & & \cdot \\ 0 & 0 & \cdot & \cdot & \cdot & \alpha_{m(1+k)-1} & 0 \\ 0 & 0 & \cdot & \cdot & \cdot & \beta_{m(1+k)} & \alpha_{m(1+k)} \end{bmatrix}$$

and

$$\nu_{M_\infty}(A) = \begin{bmatrix} \alpha_\infty & 0 & \cdot & \cdot & \cdot & 0 & \beta_\infty \\ \beta_\infty & \alpha_\infty & \cdot & \cdot & & 0 & 0 \\ \cdot & \cdot & \cdot & \cdot & \cdot & & \cdot \\ 0 & 0 & \cdot & \cdot & \cdot & \alpha_\infty & 0 \\ 0 & 0 & \cdot & \cdot & \cdot & \beta_\infty & \alpha_\infty \end{bmatrix}$$

The condition for the invertibility of $A = B_1 + B_2 W$ has the form

$$\begin{cases} \alpha_{1+mk}\alpha_{2+mk}\cdots\alpha_{m+mk} - \beta_{1+mk}\beta_{2+mk}\cdots\beta_{m+mk} \neq 0, \\ \alpha_\infty^m - \beta_\infty^m \neq 0. \end{cases}$$

Remark 23.1. If one extends the algebra $\mathcal{K}_1$ by adjoining the multiplication operators, or algebra $\mathcal{K}_2$ by adjoining the operator S, then the resulting algebras will contain all compact operators, and hence, by Corollary 22.1, will possess no sufficient family of n-dimensional representations. We shall, however, not eliminate these algebras from considerations. They are good examples of Banach algebras with a matrix symbol, and will be characterized in Sec.25.

We mention the following property of the class ω_n.

THEOREM 23.4. *Suppose that $\mathcal{K} \in \omega_n$ and let $\{\nu_M\}$ be the sufficient family of n-dimensional representations defined by formula (23.6). Then $\operatorname{Im} \nu_M = \mathbb{C}^{n \times n}$ for every $M \in \mathcal{M}$.*

PROOF. Let M_0 be an arbitrary maximal ideal of algebra $\mathcal{K}_0$. Since $p_j^2 = p_j$, $p_j(M_0)$ may assume only one of the two values: zero or one. But $\sum p_j = e$, and hence $\sum p_j(M_0) = 1$. Consequently, there is a $j \in \{1, 2, \ldots, n\}$ such that $p_j(M_0) = 1$ and $p_k(M_0) = 0$ for $k \neq j$. We must show that $\operatorname{Im} \nu_{M_0} = \mathbb{C}^{n \times n}$. To this end it suffices to show that the algebra $\operatorname{Im} \nu_{M_0}$ contains every matrix unit, i.e., every matrix which has only one entry different from zero (and equal to one). Fix m, k and choose the element $x = p_{m+i-1} V^{k-m} \in \mathcal{K}$. The entries of the matrix $\mu(x) = [x_{rq}]$ are equal to

$$x_{rq} = \sum_{j=1}^n p_j v^{r-1} p_{m+i-1} v^{k-1} v^{1-q} p_j = \sum_{j=1}^n p_j p_{m-r+i} p_{r+k-m-q+j} v^{r+k-m-q}$$

$$= \begin{cases} p_{m-r+i} & \text{if } r - q = m - k, \\ 0 & \text{if } r - q \neq m - k. \end{cases}$$

Therefore,

$$x_{rq}(M) = \begin{cases} 1 & \text{if } r = m, \quad q = k, \\ 0 & \text{for the remaining indices,} \end{cases}$$

i.e., $\nu_{M_0}(x) = [\delta_{rm}\delta_{qk}]_{r,q=1}^n$.

$\square$

We conclude the section with another example. Let $\mathcal{K}_4$ denote the Banach subalgebra of $\mathcal{L}(L(\Gamma_0))$ generated by the operators S and $(W_1\varphi)(t) = \varphi(1/t)$. In contrast to $\mathcal{K}_1$, subalgebra $\mathcal{K}_4$ is five-dimensional: $\mathcal{K}_4 = \{\alpha_1 I + \alpha_2 S + \alpha_3 W_1 + \alpha_4 S W_1 + \alpha_5 K : \alpha_j \in \mathbb{C}\}$, where

$$(K\varphi)(t) = \frac{1}{2\pi} \int_{\Gamma_0} \varphi(\tau)|d\tau|.$$

Here $W_1 S + S W_1 = 2K$.

It is known that every five-dimensional noncommutative semisimple algebra over $\mathbb{C}$ is isomorphic to the direct sum $\mathbb{C}^{2\times 2} \dotplus \mathbb{C}$, and hence belongs to π_2 [68]. A sufficient family of finite-dimensional irreducible representations consists of two homomorphisms: $\nu_1: \mathcal{K}_4 \to \mathbb{C}^{2\times 2}$ and $\nu_2: \mathcal{K}_4 \to \mathbb{C}$.

By Theorem 23.4, $\mathcal{K}_4 \notin \omega_2$. Thus, the class π_2 is wider than ω_2. However, every algebra $\mathcal{K} \in \pi_2$ belongs "locally" to ω_2 or ω_1. The class $\omega_n(\text{loc})$ is introduced in the next section.

24. LOCAL CONDITIONS

Let $\mathcal{K}$ be a Banach algebra, Δ its center, Δ_0 a Banach subalgebra of Δ, and $\mathcal{N}$ the set of all maximal ideals of Δ_0. For $N \in \mathcal{N}$ we let J_N denote the closure of the set of all finite sums of the form $\sum x_k a_k$ with $x_k \in N$ and $a_k \in \mathcal{K}$. For some ideals N it may happen that $J_N = \mathcal{K}$. For example, let $\mathcal{K} = C(\Gamma_0)$, $\Delta = \mathcal{K}$, and $\Delta_0 = C_+(\Gamma_0) =$ the set of functions which are analytic in the disc $|z| < 1$ and continuous in the closed disc $|z| \leq 1$. In this case the set $\mathcal{N}$ is homeomorphic to the closed disc $|z| \leq 1$. Take $N_0 = \{a \in C_+(\Gamma_0) : a(0) = 0\} \in \mathcal{N}$. Every function $a \in \mathcal{K}$ can be written in the form $a(t) = t(t^{-1}a)$, and hence belongs to J_{N_0}, i.e., $J_{N_0} = \mathcal{K}$. We show, however, that there exist ideals $N \in \mathcal{N}$ such that $J_N \neq \mathcal{K}$. To this end we need

LEMMA 24.1. (G.R. Allan).   *Let $\mathcal{I}$ be a left maximal ideal of the algebra $\mathcal{K}$ and let $\mathcal{A}$ be a closed commutative subalgebra (with unit) of $\mathcal{K}$. Then $\mathcal{I} \cap \mathcal{A}$ is a two-sided maximal ideal of $\mathcal{A}$.*

PROOF. $\mathcal{I} \cap \mathcal{A}$ is clearly a two-sided ideal of algebra $\mathcal{A}$. We show that it is maximal. For each $a \in \mathcal{A} \backslash \mathcal{I}$ we set

$$K_a = \{y \in \mathcal{K} : ya \in \mathcal{I}\} .$$

Then obviously K_a is a proper left ideal of $\mathcal{K}$ and $I \subset K_a$ (if $a \in \mathcal{A} \backslash \mathcal{I}$ and $y \in \mathcal{I}$, then $ya = ay \in \mathcal{I}$). Hence, $K_a = \mathcal{I}$ for every $a \in \mathcal{A} \backslash \mathcal{I}$. Consequently, if $a \in \mathcal{A} \backslash \mathcal{I}$ and the elements $y_1 a - e$ and $y_2 a - e$ belong to $\mathcal{I}$, then $y_1 - y_2 \in K_a = \mathcal{I}$. On the other hand, it follows from the maximality of the ideal $\mathcal{I}$ that for each $a \in \mathcal{A} \backslash \mathcal{I}$ there is an element $y \in \mathcal{A} \backslash \mathcal{I}$ such that $ya - e \in \mathcal{I}$. We thus showed that for every $a \in \mathcal{A} \backslash \mathcal{I}$ there is an element $y \in \mathcal{K}$, *uniquely defined modulo $\mathcal{I}$*, such that $ya - e \in \mathcal{I}$.

Our task is to show that for every element $a \in \mathcal{A} \backslash \mathcal{I}$ there is a complex number λ such that $a - \lambda e \in \mathcal{I}$. Proceeding by reductio ad absurdium, let $b \in \mathcal{A} \backslash \mathcal{I}$ be such that $b - \lambda e \in \mathcal{A} \backslash \mathcal{I}$ for all $\lambda \in \mathbb{C}$. We let $y(\lambda) \in \mathcal{K}$ denote the element for which $y(\lambda)(b - \lambda e) - e \in \mathcal{I}$.

For each point $\lambda_0 \in \mathbb{C}$ the function

$$y_1(\lambda) = y(\lambda_0)[e - (\lambda - \lambda_0)y(\lambda_0)]^{-1}$$

is holomorphic in a neighborhood $U(\lambda_0)$ of λ_0, and

$$y_1(\lambda)(b - \lambda e) = [e - (\lambda - \lambda_0)y(\lambda_0)]^{-1}[y(\lambda_0)(b - \lambda_0 e) - e] \in \mathcal{I} .$$

It follows that the classes $\widehat{y(\lambda)}$ and $\widehat{y_1(\lambda)}$ of the elements $y(\lambda)$ and $y_1(\lambda)$ in the quotient space $\mathcal{K}/\mathcal{I}$ coincide in $U(\lambda_0)$. Consequently, the function $\lambda \longrightarrow \widehat{y(\lambda)}$ is holomorphic in the complex plane $\mathbb{C}$. Since, in addition, $\widehat{y(\infty)} = 0$, it follows that $y(\lambda) \in \mathcal{I}$ for all $\lambda \in \mathbb{C}$. This leads to a contradiction: since $y(0)b - e \in \mathcal{I}$ and $\mathcal{I}$ is an ideal, $e \in \mathcal{I}$.

$\square$

Let $\mathcal{I}$ be an arbitrary left maximal ideal of algebra $\mathcal{K}$ and $N = \mathcal{I} \cap \Delta_0$. By Lemma 24.1, $N \in \mathcal{N}$ (see [3]). Let us show that $J_N \subset \mathcal{I}$, which will imply that $J_N \neq \mathcal{K}$. If $a_k \in \mathcal{K}$ and $x_k \in N$, then $\sum x_k a_k \in \mathcal{I}$. Therefore, the set $\{\sum x_k a_k\}$ is contained in $\mathcal{I}$, and hence so is its closure J_N.

Let $\mathcal{N}_0 = \{N \in \mathcal{N} : J_N \neq \mathcal{K}\}$. It is readily verified that J_N is a closed two-sided ideal of the algebra $\mathcal{K}$ for every $N \in \mathcal{N}_0$.

THEOREM 24.1. *If $a \in \mathcal{K}$ and $N \in \mathcal{N}_0$, let a_N denote the element of the quotient algebra $\mathcal{K}_N = \mathcal{K}/J_N$ which contains a. Then*

$$a \in G\mathcal{K} \Longleftrightarrow (a_N \in G\mathcal{K}_N \quad \forall N \in \mathcal{N}_0).$$

PROOF. The assertion $a \in G\mathcal{K} \Longrightarrow a_N \in G\mathcal{K}_N$ is obvious. We prove the converse by reductio ad absurdum. Suppose that $a_N \in G\mathcal{K}_N$ for all $N \in \mathcal{N}_0$, but $a \notin G\mathcal{K}$, for example, a is not left invertible. Let $\mathcal{I}_a$ denote the left maximal ideal of $\mathcal{K}$ which contains the set $K_a = \{xa : x \in \mathcal{K}\}$, and let $N_a = \mathcal{I}_a \cap \Delta_0$. By Lemma 24.1, $N_a \in \mathcal{N}$. We show that $N_a \in \mathcal{N}_0$. Since $\sum x_k a_k \in \mathcal{I}_a$ for arbitrary $x_k \in N_a$ and $a_k \in \mathcal{K}$, it follows that $J_{N_a} \subset \mathcal{I}_a \neq \mathcal{K}$. By assumption, $a_{N_a} \in G\mathcal{K}_{N_a}$, i.e., there is a $z \in \mathcal{K}$ such that $za - e \in J_{N_a} \subset \mathcal{I}_a$. Moreover, $za \in K_a \subset \mathcal{I}_a$, and hence $e \in \mathcal{I}_a$, which is impossible.

$\square$

Henceforth we will write $\mathcal{K} \in \tilde{\omega}_n$ whenever $\mathcal{K}/\mathcal{R}(\mathcal{K}) \in \omega_n$. It follows from Lemma 22.1 and Theorem 23.3 that

$$\mathcal{K} \in \tilde{\omega}_n \Longrightarrow \mathcal{K} \in \pi_n. \tag{24.1}$$

Definition. We say that the algebra $\mathcal{K}$ *belongs to* ω_n *locally*, and write $\mathcal{K} \in \omega_n(\text{loc})$, if $\mathcal{K}_N \in \tilde{\omega}_\ell$, with $\ell = \ell(N) \le n$, for every ideal $N \in \mathcal{N}_0$. In particular, if $\mathcal{K}_N \in \omega_\ell$ $(\ell = \ell(N) \le n)$ for every $N \in \mathcal{N}_0$, then $\mathcal{K} \in \omega_n(\text{loc})$.

THEOREM 24.2. *If* $\mathcal{K} \in \omega_n(\text{loc})$, *then* $\mathcal{K} \in \pi_n$.

PROOF. By assumption, for every $N \in \mathcal{N}_0$ the quotient algebra $\mathcal{K}_N$ belongs to $\tilde{\omega}_\ell$ $(\ell \le n)$ and hence, by (24.1), to π_n. Let $\{\nu_{M_N}\}_{M_N \in \alpha_N}$ be a sufficient family of representations of $\mathcal{K}_N$. Let α denote the set of pairs $M = (N, M_N)$ where N runs through $\mathcal{N}_0$ and, for each N, M_N runs through α_N. We set $\nu_M(x) = \nu_{M_N}(x_N)$. Then $det\ \nu_M(x) \ne 0$ for all $M \in \alpha$ if and only if $det\ \nu_{M_N}(x_N) \ne 0$ for all pairs (N, M_N), i.e., if and only if $x_N \in G\mathcal{K}_N$ for all $N \in \mathcal{N}_0$. By Theorem 24.1, the latter holds if and only if $x \in G\mathcal{K}$. We have thus shown that $\{\nu_M\}_{M \in \alpha}$ is a sufficient family of n-dimensional representations of $\mathcal{K}$, and hence that $\mathcal{K} \in \pi_n$.

□

To see how this theorem works, we apply it to the algebra $\mathcal{K}_4$, defined in the preceding section.

It is readily verified that $\Delta_0 = \{\alpha I + \lambda K \colon \alpha, \lambda \in \mathbb{C}\}$ is a subalgebra of the center of $\mathcal{K}_4 = \{\alpha I + \beta S + \gamma W + \delta SW + \lambda K\}$. Since $K^2 = K$, Δ_0 contains exactly two maximal ideals: $N_1 = \{\lambda K \colon \lambda \in \mathbb{C}\}$ and $N_2 = \{\lambda (I - K) \colon \lambda \in \mathbb{C}\}$. Following the scheme proposed above, we find the sets J_{N_1} and J_{N_2}. One can check directly that $J_{N_1} = \{\lambda K\}$ and $J_{N_2} = \{\alpha I + \beta S + \gamma W + \delta SW - (\alpha + \beta + \gamma + \delta)K\}$. The quotient algebras $\mathcal{K}_4/J_{N_2}$ and $\mathcal{K}_4/J_{N_1}$ are isomorphic to $\mathbb{C}$ and to the algebra $\mathcal{K}_1$ considered in the preceding section, respectively. Consequently, $\mathcal{K}_4/J_{N_2} \in \omega_1$, $\mathcal{K}_4/J_{N_1} \in \omega_2$, and hence $\mathcal{K}_4 \in \omega_2(\text{loc})$. Moreover, for $A = \alpha I + \beta S + \gamma W + \delta SW + \lambda K \in \mathcal{K}_4$,

$$\nu_1(A) = \begin{bmatrix} \alpha + \beta & \gamma + \delta \\ \gamma - \delta & \alpha - \beta \end{bmatrix}$$

and

$$\nu_2(A) = \alpha + \beta + \gamma + \delta + \gamma.$$

Hence, the operator A is invertible if and only if

$$\alpha^2 - \beta^2 - \gamma^2 + \delta^2 \neq 0 \quad \text{and} \quad \alpha + \beta + \gamma + \delta + \lambda \neq 0. \tag{24.2}$$

We return to our investigation of the classes $\omega_n(\text{loc})$. Theorem 24.2 and Lemma 22.1 admit

COROLLARY 24.1. *If $\mathcal{K}/\mathcal{R}(\mathcal{K}) \in \omega_n(\text{loc})$, then $\mathcal{K} \in \pi_n$.*

$\square$

For $n \leq 2$ the converse is also valid:

THEOREM 24.4. *Let $n \leq 2$. If $\mathcal{K} \in \pi_n$, then $\mathcal{K}/\mathcal{R}(\mathcal{K}) \in \omega_n(\text{loc})$.*

PROOF. For $n = 1$ this assertion follows from Theorem 13.1. To prove it for $n = 2$ we need three lemmas.

LEMMA 24.2. *Let $\mathcal{K}$ be a semisimple algebra and Δ its center. If $z \in \Delta$ and $N \in \mathcal{N}_0$, then z_N is a scalar element: $z_N = \lambda_N e_N$.*

PROOF. Let $\lambda_N = z(N)$ be the value of the Gelfand transform of z on the ideal N. Then $(z - \lambda_N e_N)(N) = 0$. Consequently, $z - \lambda_N e \in N \subset J_N$, and hence $z_N = \lambda_N e_N$.

$\square$

LEMMA 24.3. *Let $\mathcal{K}$ be a noncommutative semisimple algebra. If $[x, y]^2$ is a scalar element for all $x, y \in \mathcal{K}$, then $\mathcal{K} \in \omega_2$.*

PROOF. We first show that there are elements $a, b \in \mathcal{K}$ such that $[a, b]^2 = 0$. Suppose that $[a, b]^2 = 0$ for all $a, b \in \mathcal{K}$. Let x, a, b be arbitrary fixed elements. Then $0 = [x, [a, b]]^2 = x[a, b]x[a, b] - [a, b]x^2[a, b] + [a, b]x[a, b]x$. Multiplying both sides of this equality by $x[a, b]$ on the left we obtain $0 = (x[a, b])^3 \implies e + x[a, b] \in G\mathcal{K} \implies [a, b] \in \mathcal{R}(\mathcal{K}) \implies [a, b] = 0$; contradiction.

Therefore, there exist $u, w \in \mathcal{K}$ such that $[u, w]^2 \neq 0$. But $[u, w]^2 = \lambda e$ and since $\mathcal{K}$ is an algebra over the complex field $\mathbb{C}$, we may assume that $[u, w]^2 = e$. Set $[u, w] = s$, $p = \frac{1}{2}(e + s)$, and $q = \frac{1}{2}(e - s)$. Since $s^2 = e$, it follows that $p^2 = p$, $q^2 = q$,

$pq = qp = 0$, and $p + q = e$. We show that there is an element v such that $sv + vs = 0$ and $v^2 = e$. This will imply that $v^{-1}pv = q$ and $v^{-1}qv = p$.

Let us show, by reductio ad absurdum, that there is a c such that $[c, s]^2 \neq 0$. Thus, suppose that $[z, s]^2 = 0$ for all $z \in \mathcal{K}$. Let $a, b \in \mathcal{K}$ and $z = pap + qbq$. Since $ps = sp = p$ and $qs = sq = -q$,

$$[z, s] = 2(qbp - paq)$$

and

$$0 = [z, s]^2 = -4(qbpaq + paqbp).$$

Multiplying this equality on the left by p, and then by q, we get

$$qbpaq = 0, \qquad paqbp = 0. \tag{24.3}$$

We show that this implies $paq = 0$ and $qbp = 0$. It suffices to show that $paq \in \mathcal{R}(\mathcal{K})$ and $qbp \in \mathcal{R}(\mathcal{K})$. Let $x \in \mathcal{K}$. Then $(xpaq)^2 = (pxpaq + qxpaq)^2 = (pxpaq)^2 = 0$, and hence $paq \in \mathcal{R}(\mathcal{K})$. Similarly, $qbp \in \mathcal{R}(\mathcal{K})$. Consequently, $paq = qbp = 0$. This implies that $p \in \Delta(\mathcal{K})$, the center of $\mathcal{K}$. In fact, $ap = pap + qap = pap = pa - paq = pa$ for every $a \in \mathcal{K}$. Now, since $s = 2p - e$, this shows that $s \in \Delta$, and hence that $e = s^2 = [su, w]$, which is impossible.

We have thus shown that there is a $c \in \mathcal{K}$ such that $[c, s]^2 \neq 0$. We may assume that $[c, s]^2 = e$ (recall that $[c, s]^2 = \lambda e$). Set $v = [c, s]$. Since $v^2 = e$ and $vs + sv = 0$, v is the desired element.

To complete the proof of the lemma, we show that the subalgebra $\mathcal{K}_0 = p\mathcal{K}p + q\mathcal{K}q$ is commutative. To this end we first of all remark that $\xi = [x, y][z, y] + [z, y][x, y]$ is a scalar element for every choice of $x, y, z \in \mathcal{K}$. This follows from the fact that in the equality $[x + z, y]^2 = [x, y]^2 + [z, y]^2 + \xi$ we have $[x + z, y]^2 = \lambda_1 e$, $[x, y]^2 = \lambda_2 e$, and $[z, y]^2 = \lambda_3 e$, with $\lambda_1, \lambda_2, \lambda_3 \in \mathbb{C}$.

Now let $a, b \in \mathcal{K}$, and set $x = qbpv$, $y = pv$, and $z = vp$. Then $[x, y] = qbq - pvbvp$, $[z, y] = -s$, and $\xi = 2(qbp + pvbvp)$. Let $\eta = qaq$. Since $\xi p = p\xi$,

$$qbpap = qaqbq. \tag{24.4}$$

Similarly, the choice $x = pfpv$, $y = qv$, $z = vq$ and $\eta = pgp$, with $f, g \in \mathcal{K}$, yields

$$pfpgp = pgpfp. \tag{24.5}$$

It follows from (24.4) and (24.5) that algebra $\mathcal{K}_0$ is commutative.

□

LEMMA 24.4. *If* $[[x,y]^2, z] \in \mathcal{R}(\mathcal{K})$ *for all* $x, y, z \in \mathcal{K}$, *then* $\mathcal{K}/\mathcal{R}(\mathcal{K}) \in \omega_2(\mathrm{loc})$.

PROOF. The assumption of the lemma implies that $[[a,b]^2, c] = 0$ for all $a, b, c \in \mathcal{K}/\mathcal{R}(\mathcal{K})$. By Lemma 24.2, $[a_N, b_N]^2 = \lambda_N e_N$ is a scalar element in $(\mathcal{K}/\mathcal{R}(\mathcal{K}))_N$ (here the ideal J_N is constructed in the algebra $\mathcal{K}/\mathcal{R}(\mathcal{K})$). Consequently, $[a_N, b_N]^2 = \lambda_N e_N$ for all a_N and b_N. By Lemma 24.3, $\mathcal{K}_N \in \tilde{\omega}_k$ $(k = 1, 2)$ for every $N \in \mathcal{N}_0$, and hence $\mathcal{K}/\mathcal{R}(\mathcal{K}) \in \omega_2(\mathrm{loc})$.

□

CONTINUATION OF THE PROOF OF THEOREM 24.4. We can take the center Δ itself for a subalgebra of Δ. Suppose that $\mathcal{K} \in \pi_2$. Since $[[x,y]^2, z] = 0$ for all $x, y, z \in \mathbb{C}^{2 \times 2}$, it follows that $[[x,y]^2, z] \in \mathcal{R}(\mathcal{K})$ for all $x, y, z \in \mathcal{K}$. Now apply Lemma 24.4.

□

We shall write $\mathcal{K} \in \pi$ whenever the algebra $\mathcal{K}$ possesses a sufficient family of finite-dimensional representations (with no constraints on the dimensionalities of the representations).

Theorem 24.1 admits the following corollary.

COROLLARY 24.2. *Suppose that* $\mathcal{K}_N \in \pi_{n(N)}$ *(or* $\omega_{n(N)}$*) for every* $N \in \mathcal{N}_0$. *Then* $\mathcal{K} \in \pi$.

□

We consider an example. Let $E = \ell_2$, $\mathcal{K}$ the subalgebra of $\mathcal{L}(E)$ generated by the operators B, $B\{\xi_n\} = \{\lambda_n \xi_n\}$, of multiplication by sequences of complex numbers λ_n for which the limit $\lambda_\infty = \lim_{n \to \infty} \lambda_n$ exists, the shift operator W, where $W\{\xi_n\} = \{\xi_{\sigma(n)}\}$, and its inverse W^{-1}. Here $\sigma \colon \mathbf{N} \to \mathbf{N}$ is defined by the formulas $\sigma(n(n+1)/2) = n(n+1)/2 + 1 - n$, $\sigma(k) = k + 1$ for $k \neq n(n+1)/2$.

Notice that in the algebra $\mathcal{K}_3$ considered in the preceding section the operator

W enjoyed the property that $W^m = I$. In the present example $W^k \neq I$ for all $k \neq 0$. Such an operator is usually referred to as a *non-Carleman shift* [53].

Let us show that $\mathcal{K} \notin \pi_n$ for any n. Since $\mathcal{K}$ is a self-adjoint subalgebra of $\mathcal{L}(\ell_2)$, it is semisimple. If $\mathcal{K}$ were to belong to some class π_m, then by Theorem 22.1, the standard identity $F_{2m}(a_1, \ldots, a_{2m}) = 0$ would hold on $\mathcal{K}$. We write the operators $A \in \mathcal{K}$ in matrix form relative to the standard basis of ℓ_2. Then every A has the block-diagonal form $A = \mathrm{diag}(c_1, c_2, \ldots, c_n, \ldots)$, where $c_n \in \mathbb{C}^{n \times n}$. It is readily verified that the mapping $A \to c_n$ is an epimorphism of the algebra $\mathcal{K}$ onto $\mathbb{C}^{n \times n}$. Since the identity $F_{2m}(a_1, \ldots, a_{2m}) = 0$ is not fulfilled on $\mathbb{C}^{n \times n}$ for $m < n$, we reach a contradiction. Therefore, $\mathcal{K} \notin \pi_n$. Nevertheless, $\mathcal{K}_N \in \pi_{n(N)}$ for every $N \in \mathcal{N}_0$. It is easily seen that the set Δ_0 of all operators of multiplications by sequences $\{\lambda_n\}$ with the property that $\lambda_{m+1} = \lambda_{m+2} = \ldots = \lambda_{m+k}$ (where $2m = (k+1)k - 2k$ and $k \in \mathbb{N}$) is a subalgebra of the center of $\mathcal{K}$.

For each natural number k the set N_k of operators from Δ_0 for which $\lambda_{m+1} = \lambda_{m+2} = \ldots = \lambda_{m+k} = 0$ (where $m = k(k-1)$) is a maximal ideal of Δ_0. The set N_∞ of all operators from Δ_0 for which $\lambda_\infty = 0$ is also a maximal ideal, and it is readily verified that this is a complete list of the maximal ideals of Δ_0. For each maximal ideal N_k (N_∞) there is the corresponding ideal J_{N_k} (respectively, J_{N_∞}). J_{N_k} consists of all operators $A \in \mathcal{K}$ in the block-diagonal representation $A = \mathrm{diag}(A_1, A_2, \ldots)$ of which the block $A_k = 0$. Consequently, the quotient algebra $\mathcal{K}_{N_k}$ is isomorphic to $\mathbb{C}^{k \times k}$, and the corresponding representation is $\nu_k(A) = A_k$. The quotient algebra $\mathcal{K}_{N_\infty}$ is isomorphic to the commutative algebra $\overset{\circ}{\mathcal{K}}$ generated by the operators W and W^{-1}. It is readily checked that the natural homomorphism $f \colon \overset{\circ}{\mathcal{K}} \to \mathcal{K}_{N_\infty}$ is an isomorphism.

Thus, $\mathcal{K}_{N_k} \in \pi_k$, $\mathcal{K}_{N_\infty} \in \pi_1$, and by Corollary 24.2, $\mathcal{K} \in \pi$.

In order to find a sufficient family of finite-dimensional representations it remains to describe the maximal ideals of the commutative C^*-algebra $\overset{\circ}{\mathcal{K}}$. The operator W is unitary, and its spectrum fills the whole circle Γ_0 (every number $exp\,(2\pi ik/n)$ is an eigenvalue of W). It follows that $\overset{\circ}{\mathcal{K}}$ is isometrically isomorphic to the algebra of all continuous functions on Γ_0. The corresponding homomorphisms are defined by the rule

$$\nu_{\infty,\tau}(A) = A_{N_\infty}(\tau) \qquad (|\tau| = 1). \tag{24.6}$$

To these we adjoin the homomorphism

$$\nu_k(A) = A_k, \qquad k = 1, 2, \ldots, \tag{24.7}$$

found previously, thereby obtaining a sufficient family of finite-dimensional homomorphisms.

In particular, for the operator $A = B_1 + B_2 W$, where $B_1\{\xi_n\} = \{\alpha_n \xi_n\}$ and
$B_2\{\xi_n\}\xi = \{\beta_n \xi_n\}$,

$$\nu_k(A) = \begin{bmatrix} \alpha_{m+1} & 0 & \cdot & \cdot & \cdot & 0 & \beta_{m+1} \\ \beta_{m+2} & \alpha_{m+2} & \cdot & \cdot & \cdot & 0 & 0 \\ \cdot & & \cdot & \cdot & \cdot & & \cdot \\ 0 & 0 & \cdot & \cdot & \cdot & \alpha_{m+k-1} & 0 \\ 0 & 0 & \cdot & \cdot & \cdot & \beta_{m+k} & \alpha_{m+k} \end{bmatrix},$$

where $2m = k^2 - k + 2$, $k = 2, 3, \ldots$, and

$$\nu_{\infty,\tau}(A) = \alpha + \beta\tau ,$$

where $\alpha = \lim \alpha_n$, $\beta = \lim \beta_n$, and $\tau \in \Gamma_0$. Hence, A is invertible if and only if

$$\alpha_{m+1}\alpha_{m+2}\cdots\alpha_{m+k} + (-1)^k \beta_{m+1}\beta_{m+2}\cdots\beta_{m+k} \neq 0 \qquad (k = 1, 2, \ldots, 2m = k^2 - k + 2)$$

and $\alpha + \beta\tau \neq 0$, $|\tau| = 1$.

<u>Remark</u>. Since $\mathcal{K}$ is a C^*-subalgebra of $\mathcal{L}(\ell_2)$, an operator $A \in \mathcal{K}$ is invertible
in $\mathcal{L}(\ell_2)$ if and only if it is invertible in $\mathcal{K}$.

25. A TOPOLOGY ON THE SET OF MAXIMAL IDEALS OF BANACH PI-ALGEBRAS

I. Let $\mathcal{K}$ be a Banach PI-algebra and $\mathcal{M}$ the set of all its two-sided maximal ideals. If $\mathcal{K}$ is semisimple, then it is isomorphic to the algebra of functions $\varphi\colon \mathcal{M} \to \mathbb{C}^{\ell \times \ell}$ where $\ell = \ell(M)$, $M \in \mathcal{M}$. We define a topology on $\mathcal{M}$. To this end we set $\mathcal{M}_k = \{M \in \mathcal{M}\colon \ell(M) = k\}$, and then for every $\varepsilon > 0$ and arbitrary $x_1, \ldots, x_r \in \mathcal{K}$ and $M_0 \in \mathcal{M}_k$,

$$U_{x_1,\ldots,x_r,\varepsilon}(M_0) = \{M \in \mathcal{M}_k\colon \|\nu_M(x_k) - \nu_{M_0}(x_k)\| < \varepsilon, \quad k = 1, \ldots, r\}, \tag{25.1}$$

where $\{\nu_M\}_{M \in \mathcal{M}}$ is a sufficient family of n-dimensional representations, and the norm of a matrix is defined as the norm of the corresponding operator acting in $\mathbb{C}^k$. It is readily verified that the sets (25.1) can serve as a basis of neighborhoods of the point M_0 in a certain topology that we will call the *Gelfand topology* on $\mathcal{M}$. It is Hausdorff, and for every $x \in \mathcal{K}$ the function $\varphi_x(M) = \nu_M(x)$ is continuous on $\mathcal{M}$. However, $\mathcal{M}$ is not necessarily compact.

Example 25.1. Let $\mathcal{K}$ denote the set of all matrix-valued functions $f\colon [0,1] \to \mathbb{C}^{2 \times 2}$, $f = [f_{jk}]_{j,k=1}^2$, for which $f_{12}(0) = f_{21}(0) = 0$. In this example $\mathcal{M} = \mathcal{M}_1 \cup \mathcal{M}_2$, where $\mathcal{M}_2$ is homeomorphic to the semiinterval $(0,1] \subset \mathbb{R}$, and $\mathcal{M}_1$ is a two-point space. Hence, $\mathcal{M}$ is not compact.

If there is given a sufficient family of representations on $\mathcal{K}$, then each representation can be replaced by an equivalent one without destroying the "sufficiency" of the family. Thanks to this we can obtain families in which the representations enjoy certain supplementary properties. We shall need the following lemma:

LEMMA 25.1. *Let $\mathcal{K}$ be a Banach PI-algebra. Then there exists a sufficient family $\{\nu_M\}_{M \in \mathcal{M}}$ of finite-dimensional irreducible representations $\nu_M\colon \mathcal{K} \to \mathbb{C}^{\ell \times \ell}$, with $\ell = \ell(M) \le n$, such that $\operatorname{Im} \nu_M = \mathbb{C}^{\ell(M) \times \ell(M)}$ and the matrices $\nu_M(x) = [x_{jk}(M)]$ satisfy the bound*

$$|x_{jk}(M)| \le \|x\|.$$

PROOF. Let ξ_M be the left regular representation of the algebra $\mathcal{K}_M = \mathcal{K}/M$. In Sec.21 we have shown that $Im\ \xi_M = \mathcal{L}(E_M)$, where $dim\ E_M \leq n$, and that $\|\xi_M\| = 1$. In the space E_M we choose an Auerbach basis, i.e., a basis $\{z_j\}_{j=1}^{\ell}$ with the property that if $f_i \in E_M^*$ are functionals such that $f_k(x_j) = \delta_{kj}$ (where δ_{kj} is the Kronecker symbol) then $\|z_j\| = \|f_j\| = 1$ (see[36]). Let $\mu_M(x)$ denote the matrix of the operator $A_M = \xi_M(x_M)$ in the basis $\{z_j\}$. Then $|x_{jk}(M)| = |f_j(A_M z_k)| \leq \|\xi_M(x_M)\| \leq \|\xi_M\| \cdot \|x_M\| \leq \|x\|$, as claimed.

$\square$

Throughout this section we shall assume that the family $\{\nu_M\}$ has properties analogous to those of the family whose existence was established in Lemma 25.1.

II. Let us examine more closely the case where $Im\ \nu_M = \mathbb{C}^{n \times n}$ for every $M \in \mathcal{M}$. We assign to each element $x \in \mathcal{K}$ the n^2-dimensional disc $D_n(x) = D(x) \times \cdots \times D(x)$, where $D(x) = \{\lambda \in \mathbb{C} : |\lambda| \leq \|x\|\}$. Since $D_n(x)$ is compact for any n, the product $D = \prod_{x \in \mathcal{K}} D_n(x)$ is also compact, by a theorem of Tikhonov. Now to each maximal ideal $M \in \mathcal{M}$ we assign the point $\{\nu_M(x)\} \in D$. Comparing topologies it is readily seen that this rule defines a homeomorphic embedding of $\mathcal{M}$ in D. If $\mathcal{M}$ were closed in D, this would imply the compactness of $\mathcal{M}$. But in the general case $\mathcal{M}$ is not closed in D (we discuss this aspect at the end of this section).

We say that $\mathcal{K}$ satisfies *condition* α_n if $dim\ Im\ \nu_M = n$ for every $M \in \mathcal{M}$ and the set $\mathcal{M}$ is compact in D.

The above discussion proves the following result:

THEOREM 25.1. *Suppose that $\mathcal{K}$ satisfies condition α_n. Then the neighborhoods (25.1) define a compact Hausdorff topology on $\mathcal{M}$, in which the functions $x_{jk}(M)$ are continuous.*

$\square$

Let T be a compact topological space and let $\mathcal{A}$ be a C^*-algebra. We let $C(T, \mathcal{A})$ denote the algebra of all continuous $\mathcal{A}$-valued functions on T. It is endowed with a natural structure of a C^*-algebra: $f^*(t) = f(t)^*$ and $\|f\| = \max_{t \in T} \|f(t)\|$. Let B be a subset of $C(T, \mathcal{A})$. We say that B *separates the points of T* if

1) $\forall\, t_1, t_2 \in T \quad t_1 \neq t_2 \quad \exists f \in B: \quad f(t_1) = 0 \ \text{ and } \ f(t_2) = e$

and

2) $\forall\, t \in T \quad \forall a \in \mathcal{A} \quad \exists f \in B: \quad f(t) = a.$

THEOREM 25.2. *Let B be a C^*-subalgebra of $C(T, \mathcal{A})$ which separates the points of T. Then $B = C(T, \mathcal{A})$.*

PROOF. The proof is broken into a number of steps.

STEP 1. By assumption, given arbitrary distinct points $t_1, t_2 \in T$ there is an $x \in B$ such that $x(t_1) = e$ and $x(t_2) = 0$. We show that there exists a $y \in B$ such that $y(t_1) = e$ and $y(t) = 0$ in a neighborhood $U(t_2)$ of t_2. Replacing x by xx^*, we may assume that x belongs to the cone B^+ of positive elements of algebra B. Let $f: \mathbb{R}_+ \to \mathbb{R}_+$ be a continuous function which is equal to zero in a neighborhood of zero and to one on $[1, \infty)$. Then $f(x) \in C(T, \mathcal{A})$. We show that $f(x) \in B$. Since $\|x(t)\| \leq \beta$ and $x \in B^+$, the spectrum $\sigma(x)$ of x is contained in the segment $[0, \beta]$. Let p_n be a sequence of polynomials which converges uniformly to f on $[0, \beta]$. Then $p_n(x) \to f(x)$ in $C(T, \mathcal{A})$, and since $p_n(x) \in B$ we conclude that $f(x) \in B$, as claimed. Suppose that $f(s) = 0$ for $s \in [0, \alpha]$. Since $x(t_2) = 0$, there exists a neighborhood $U(t_2)$ such that $\|x(t)\| < \alpha$ for all $t \in U(t_2)$. Set $y = f(x)$. Then for $t \in U(t_2)$

$$\|p_n(x)(t)\|^2 = \|p_n(x(t))\|^2 = \max\{\lambda : \lambda \in \sigma(p_n(x(t)) \cdot p_n(x(t))^*)\}$$

$$= \max_{s \in \sigma(x(t))} |p_n(s)|^2 \leq \max_{s \in [0, \alpha]} |p_n(s)|^2 \to 0$$

as $n \to \infty$, and hence $y(t) = 0$. Moreover, it is readily checked that $y(t_1) = e$.

STEP 2. Let K be a compact subset of T and $t_1 \notin K$. We show that there exists a $z \in B^+$ such that $z(t_1) = e$ and $z(t) = 0$ on K. By Step 1, for each point $\tau \in K$ there is a neighborhood $U(\tau)$ and an element $y_\tau \in B$ such that $y_\tau(t) = 0$ for all $t \in U(\tau)$ and $y_\tau(t_1) = e$. We choose a finite cover $U(\tau_1), \ldots, U(\tau_n)$ of K and set $z = y_{\tau_1} \cdots y_{\tau_n} y_{\tau_n}^* \cdots y_{\tau_1}^*$. Then clearly z has the desired properties.

STEP 3. Let K and L be disjoint compact subsets of T. We show that there exists a $z \in B^+$ such that $z(t) = 0$ on K and $z(t) \geq e$ on L. By Step 2, for each point $\tau \in L$ there is a $z_\tau \in B^+$ such that $z_\tau(\tau) = e$ and $z_\tau(t) = 0$ for all $t \in K$. Then $z_\tau(t) \geq e/2$

for all t in a neighborhood U_τ of τ. We choose a finite cover $U_{\tau_1}, \ldots, U_{\tau_n}$ of L and put $z = 2(z_{\tau_1} + \cdots + z_{\tau_n})$. Then $z \in B^+$, $z(t) = 0$ for all $t \in K$, and $z(t) \geq e$ for all $t \in L$.

STEP 4. Let $f : T \to \mathbb{C}$ be a continuous function and $\varepsilon > 0$. We claim that if $x \in B$ and $y(t) = f(t)x(t)$, then $y \in B$. In fact, let $W_1, \ldots, W_n$ be an open cover of T such that the oscillation of f on each W_j does not exceed ε. Let $W_1', \ldots, W_n'$ be an open cover subordinate to $W_1, \ldots, W_n$, i.e., $\bar{W}_j' \subset W_j$. Then for each $j = 1, \ldots, n$ there is a $v_j \in B^+$ such that $v_j(t) \geq e$ for all $t \in \bar{W}_j'$, whereas $v_j(t) = 0$ on $T \backslash W_j$. Set $v = v_1 + \cdots + v_n$. Since $v(t) \geq e$ on T, there is a $\tilde{v} \in B^+$ such that $\tilde{v}^2 v = v\tilde{v}^2 = e$. Set $\zeta_j = \tilde{v} v_j \tilde{v}$. Then $\sum \zeta_j = \tilde{v} v \tilde{v} = e$. Finally, put $\lambda_j = f(\tau_j)$, where $\tau_j \in W_j$. Then $f(t)e(t) - \sum \lambda_j \zeta_j(t) = f(t)\sum \zeta_j(t) - \sum \lambda_j \zeta_j(t) = \sum (f(t) - \lambda_j)\zeta_j(t)$, whence $|\sum (f(t) - \lambda_j)\zeta_j(t)| \leq \sum |f(t) - \lambda_j||\zeta_j(t)| < \varepsilon \sum \zeta_j(t) = \varepsilon e(t)$. Consequently, $\|f(t)e(t) - \sum \lambda_j \zeta_j(t)\| \leq \varepsilon$ for all $t \in T$. Therefore, $fe \in B$ and hence $fx \in B$ for every $x \in B$.

STEP 5 is to complete the proof. Let $x \in C(T, \mathcal{A})$ and $\varepsilon > 0$. Since B separates the points of T, for every $\tau \in T$ there is an $x_\tau \in B$ such that $x_\tau(\tau) = x(\tau)$, and hence $\|x_\tau(t) - x(t)\| \leq \varepsilon$ in a neighborhood $U(\tau)$ of τ. Choose a finite cover $U(\tau_1), \ldots, U(\tau_n)$ of T, and let $\eta_1, \ldots, \eta_n$, $\eta_k(t) \geq 0$ on T, be a continuous partition of unity subordinate to this cover. Then $\|x(t) - \sum \eta_k(t)x_k(t)\| = \|\sum \eta_k(t)(x - x_k)(t)\| \leq \sum \eta_k(t)\|x(t) - x_k(t)\| \leq \varepsilon$, which shows that $x \in B$.

$\square$

COROLLARY 25.1. *Let B be a C^*-subalgebra of $C(T)^{n \times n}$ with the property that for any $t_1, t_2 \in T$, $t_1 \neq t_2$, and any $R_1, R_2 \in \mathbb{C}^{n \times n}$, there is an $x \in B$ such that $x(t_1) = R_1$ and $x(t_2) = R_2$. Then $B = C(T)^{n \times n}$.*

$\square$

This is a matrix analog of the Stone-Weierstrass theorem.

II. Here we shall establish a matrix analog of the Gelfand-Naimark theorem which asserts that every commutative Banach C^*-algebra is isometrically isomorphic to the algebra of all continuous functions on the compactum of its maximal ideals.

First we prove a number of auxiliary propositions.

LEMMA 25.2. *Every finite-dimensional irreducible representation is uniquely determined (to within equivalence) by its kernel.*

PROOF. Let $\nu_1: \mathcal{A} \to \mathcal{L}(E_1)$ and $\nu_2: \mathcal{A} \to \mathcal{L}(E_2)$ be irreducible representations of the algebra $\mathcal{A}$ with $dim\ E_k < \infty$, $k = 1, 2$, and suppose that $Ker\ \nu_1 = Ker\ \nu_2$. By the basic theorem on homomorphisms, $Im\ \nu_1 \cong \mathcal{A}/Ker\ \nu_1 = \mathcal{A}/Ker\ \nu_2 \cong Im\ \nu_2$. By Burnside's theorem [80], $Im\ \nu_k = \mathcal{L}(E_k)$, $k = 1, 2$, and hence $\mathcal{L}(E_1) \cong \mathcal{L}(E_2)$; we denote this isomorphism by $\varphi: \mathcal{L}(E_1) \to \mathcal{L}(E_2)$. Then there exists a matrix $A \in G\ \mathbb{C}^{n \times n}$, where $n = dim\ E_k$, such that $\varphi(\nu_1(x)) = A^{-1}\nu_1(x)A$ (see [27]), and hence $\nu_2 = \varphi\nu_1 \cong \nu_1$. $\square$

Remark 25.1. Let $\mathcal{A}$ be a Banach PI-algebra. It follows from Lemma 25.2 that the family $\{\nu_M\}_{M \in \mathcal{M}}$ contains (to within equivalence) all irreducible representations of $\mathcal{A}$. In fact, if ν is an arbitrary irreducible representation of $\mathcal{A}$, then $M = Ker\ \nu$ is a primitive ideal. By Lemma 21.6, $\mathcal{A}/M \cong \mathbb{C}^{n \times n}$ for some value of n. Consequently, $M \in \mathcal{M}$ and ν is a finite-dimensional representation, which in turn implies that $\nu \cong \nu_M$.

LEMMA 25.3. *Let $\nu_1: \mathcal{A} \to \mathcal{L}(E_1)$ and $\nu_2: \mathcal{A} \to \mathcal{L}(E_2)$ be nonequivalent finite-dimensional irreducible representations of the algebra $\mathcal{A}$, and let $R_1 \in \mathcal{L}(E_1)$ and $R_2 \in \mathcal{L}(E_2)$. Then there exists an $x \in \mathcal{A}$ such that $\nu_k(x) = R_k$, $k = 1, 2$.*

PROOF. By Burnside's theorem, there exist elements $y, v \in \mathcal{A}$ such that $\nu_1(y) = R_1$ and $\nu_2(v) = R_2$. Set $z = y - v$, $M_1 = Ker\ \nu_1$, and $M_2 = Ker\ \nu_2$. Since $\nu_1 \not\cong \nu_2$, $M_1 \neq M_2$. But $M_1, M_2 \in \mathcal{M}$, so that $M_1 + M_2 = \mathcal{A}$. Therefore, there are elements $y_1 \in M_1$ and $v_1 \in M_2$ such that $y_1 + v_1 = z$. Now set $x = y - y_1 = v + v_1$. Then $\nu_1(x) = R_1$ and $\nu_2(x) = R_2$. $\square$

THEOREM 25.3. *Suppose that $\mathcal{A}$ satisfies condition α_n. If $\mathcal{A}$ is semisimple, then it is isomorphic to a subalgebra of $C(\mathcal{M})^{n \times n}$.*

PROOF. We show that the kernel of the homomorphism exhibited above is contained in the radical $\mathcal{R}(\mathcal{A})$ of $\mathcal{A}$. Let $x \in \mathcal{A}$ and $\hat{x}(M) = 0$ for every $M \in \mathcal{M}$. Then $\nu_M(e + ux) = \nu_M(e) + \nu_M(u)\nu_M(x) = \nu_M(e) \in G\ \mathbb{C}^{n \times n}$ for all $u \in \mathcal{A}$, and hence $x \in \mathcal{R}(\mathcal{A})$. But $\mathcal{R}(\mathcal{A}) = \{0\}$. $\square$

THEOREM 25.4. *Suppose that the C^*-algebra $\mathcal{A}$ satisfies condition α_n. Then $\mathcal{A}$*

is isometrically *-isomorphic to the algebra of all continuous $n \times n$ matrix-valued functions
on the compact space $\mathcal{M}$ of all two-sided maximal ideals of $\mathcal{A}$.

PROOF. $\mathcal{A}$ is semisimple, and hence, by Theorem 25.3, it imbeds isomorphi-
cally into the algebra $\mathcal{A}_1 = C(\mathcal{M})^{n \times n}$. Let $\mathcal{A}_2$ denote the image of $\mathcal{A}$ under this imbedding.
By Lemma 25.3, $\mathcal{A}_2$ separates the points of $\mathcal{M}$. Moreover, $\mathcal{A}_2$ is a C^*-subalgebra of $\mathcal{A}_1$.
It remains to apply Corollary 25.1.

□

We shall now consider a weaker topology on the set $\mathcal{M}$ of maximal ideals of
the Banach PI-algebra $\mathcal{A}$ under the restrictive assumption that $dim\ Im\ \nu_M = n$ for all
$M \in \mathcal{M}$.

In this topology a net $\{M_\alpha\} \subset \mathcal{M}$ converges to $M \in \mathcal{M}$ if and only if there
exist matrix representations τ_{M_α} and τ_M such that $Ker\ \tau_{M_\alpha} = M_\alpha$, $Ker\ \tau_M = M$, and
$(\tau_{M_\alpha}(x))_{j,k} \to (\tau_M(x))_{j,k}$, $j, k = 1, \ldots, n$, for all $x \in \mathcal{A}$. This coincides with the topology
introduced by J. Fell [1(5)] on the set of equivalence classes of irreducible representations
of Banach algebras for which every irreducible representation is n-dimensional. In [76(1)]
Fell showed that $\mathcal{M}$ is compact in this topology and that the C^*-algebra $\mathcal{A}$ is isomorphic to
the algebra of all continuous sections of a certain bundle with base $\mathcal{M}$ and fiber $\mathbb{C}^{n \times n}$. In
[76] the homotopy classification of fiber bundles is used to build examples of nonisomorphic
C^*-algebras for which all irreducible representations are n-dimensional (with $n > 1$) and
the representation spaces are homeomorphic. From this it follows, in particular, that for
C^*-PI-algebras with the property that $dim\ Im\ \nu_M = n$ for all $M \in \mathcal{M}$ the set $\mathcal{M}$ cannot be
closed in D. In fact, let $\mathcal{A}$ be a C^*-PI-algebra which is not isomorphic to $C(\mathcal{M})^{n \times n}$ and for
which $dim\ Im\ \nu_M = n$ $\forall M \in \mathcal{M}$. Suppose that for this algebra $\mathcal{M}$ is closed in D. Then,
by Theorem 25.4, $\mathcal{A}$ is isomorphic to $C(\tilde{\mathcal{M}})^{n \times n}$, where $\tilde{\mathcal{M}}$ designates the set $\mathcal{M}$ endowed
with the Gelfand topology (whereas $\mathcal{M}$ is endowed with the weaker, Fell topology). Since
$\mathcal{M}$ and $\tilde{\mathcal{M}}$ are both compact (and Hausdorff) we have reached a contradiction.

CHAPTER VII

MATRIX SYMBOLS

In this chapter we obtain conditions for the existence of an n-symbol and construct such a symbol for certain classes of Banach algebras. We also find conditions under which the order of the symbol cannot be reduced and then we give examples of algebras which have a matrix symbol and of those which do not.

26. CONDITIONS FOR THE EXISTENCE OF MATRIX SYMBOLS

We shall assume that $\mathcal{K}$ is a subalgebra of $\mathcal{L}(E)$, where E is a Banach space. We say that $\mathcal{K}$ is an **algebra with n-symbol (n-semisymbol)** and write $\mathcal{K} \in \sigma_n$ (respectively, $\mathcal{K} \in \sigma_n^0$), if there exists a family $\{\gamma_M\}_{M \in \alpha}$ of homomorphisms $\gamma_M : \mathcal{K} \to \mathbb{C}^{\ell \times \ell}$, with $\ell = \ell(M) \leq n$, such that

$$A \in \Phi(E) \Longleftrightarrow (det\ \gamma_M(A) \neq 0 \quad \forall M \in \alpha) \tag{26.1}$$

(respectively

$$A \in \Phi(E) \Longleftarrow (det\ \gamma_M(A) \neq 0 \quad \forall M \in \alpha)). \tag{26.2}$$

As in the previous chapters, if $\mathcal{T}(E) \subset \mathcal{K}$ we let $\hat{\mathcal{K}}$ denote the quotient algebra $\mathcal{K}/\mathcal{T}(E)$, whereas if $\mathcal{T}(E) \not\subset \mathcal{K}$ we put $\hat{\mathcal{K}} = \{\hat{A} \in \widehat{\mathcal{L}(E)} = \mathcal{L}(E)/\mathcal{T}(E) : \hat{A} \cap \mathcal{K} \neq \emptyset\}$.

Recall that algebra $\hat{\mathcal{K}}$ is said to be inverse-closed in $\widehat{\mathcal{L}(E)}$ if $\hat{A} \in \hat{\mathcal{K}} \cap G\widehat{\mathcal{L}(E)} \Longrightarrow \hat{A} \in G\hat{\mathcal{K}}$.

THEOREM 26.1. *Suppose $\hat{\mathcal{K}} \in \sigma_n$ and let $\{\nu_M\}_{M \in \alpha}$ be a sufficient family of n-dimensional representations of $\hat{\mathcal{K}}$. Then the rule $\gamma_M(A) = \nu_M(\hat{A})$ defines an n-semisymbol on algebra $\mathcal{K}$. This semisymbol is an n-symbol if and only if $\hat{\mathcal{K}}$ is inverse-closed in $\widehat{\mathcal{L}(E)}$.*

PROOF. The proof is analogous to that of Theorem 14.1.

□

In the case where $\hat{\mathcal{K}} \in \omega_n$ (see Section 23), we define the family $\{\gamma_M\}_{M \in \alpha}$ by the rule: $\gamma_M(A) = [\hat{A}_{jk}(M)]$, where $[\hat{A}_{jk}] = \mu(\hat{A})$, and $\hat{A}_{jk}(M)$ are multiplicative functionals on the commutative algebra $\hat{\mathcal{K}}_0 = \sum p_j \hat{\mathcal{K}} p_j$. The family $\{\gamma_M\}$ is said to be *symmetric* if for every $A \in \mathcal{K}$ there is an $\bar{A} \in \mathcal{K}$ such that $\gamma_M(\bar{A}) = \gamma_M(A)^*$ for all M.

THEOREM 26.2. *Let $\hat{\mathcal{K}} \in \omega_n$ and let $\mathcal{M}$ denote the space of maximal ideals of the algebra $\hat{\mathcal{K}}_0$. Then:*

1) *there exists a subset $\mathcal{M}_0 \subset \mathcal{M}$ such that the family $\{\gamma_M\}_{M \in \mathcal{M}_0}$ defines a symbol on $\mathcal{K}$;*

2) *if the family $\{\gamma_M\}_{M \in \mathcal{M}}$ is symmetric, then one can take $\mathcal{M}_0 = \mathcal{M}$;*

3) *if $A \in \mathcal{K} \cap \Phi(E)$, then $Ind\ A = \dfrac{1}{n} Ind\ [A_{jk}]$, where $A_{jk} \in \hat{A}_{jk}$;*

4) *if the set $\mathcal{K} \cap \Phi(E)$ is dense in $\mathcal{K}$, then*

$$Ind\ A = \frac{1}{n} Ind\ det\ A. \tag{26.3}$$

PROOF. 1) Let $\hat{\mathcal{L}}_0 = \sum p_j \widehat{\mathcal{L}(E)} p_j$ and let $\mathcal{K}_{max}$ denote the maximal computative subalgebra of $\hat{\mathcal{L}}_0$ which contains $\hat{\mathcal{K}}_0$. We show that

$$x \in G\hat{\mathcal{L}}_0^{n \times n} \cap \hat{\mathcal{K}}_0^{n \times n} \Longrightarrow x^{-1} \in \mathcal{K}_{max}^{n \times n} \tag{26.4}$$

Let $x = [x_{jk}]$, $x^{-1} = [y_{jk}]$, c an arbitrary element of $\mathcal{K}_{max}$, and $\tilde{c} = [\delta_{jk}c]$. It follows from the equalities $[cy_{jk}] = x^{-1}x\tilde{c}x^{-1} = x^{-1}cxx^{-1} = [y_{jk}c]$ that the entries y_{jk} commute with every element $c \in \mathcal{K}_{max}$, which proves (26.4).

By (26.4) and Theorem 23.2,

$$A \in \mathcal{K} \cap \Phi(E) \Longleftrightarrow \hat{A} \in G\widehat{\mathcal{L}(E)} \cap \hat{\mathcal{K}} \Longleftrightarrow \mu(\hat{A}) \in G\hat{\mathcal{L}}_0^{n \times n} \cap \hat{\mathcal{K}}_0^{n \times n} \Longleftrightarrow \mu(\hat{A}) \in G\mathcal{K}_{max}^{n \times n}$$

$$\Longleftrightarrow (det\ [\hat{A}_{jk}(M)] \neq 0 \quad \forall\ M \in \mathcal{M}_0),$$

where $\mathcal{M}_0$ designates the set of maximal ideals of the algebra $\mathcal{K}_{max}$. Therefore, from the semisymbol $\gamma_M(A) = [A_{ik}(M)]$, $M \in \mathcal{M}$, we must retain only those γ_M for which $x(M)$, where $x \in \hat{\mathcal{K}}_0$, extends to a multiplicative functional on $\mathcal{K}_{max}$.

2) It suffices to show that $\hat{A} \in G\widehat{\mathcal{L}(E)} \cap \hat{\mathcal{K}} \Longrightarrow \hat{A}^{-1} \in \hat{\mathcal{K}}$.

Let $B \in \mathcal{K}$ such that $[\hat{B}_{jk}(M)] = [\hat{A}_{jk}(M)]^*$. Let $a = det\ \mu(\hat{A})$, $b = det\ \mu(\hat{B})$, $x = (a+b)/2$, and $y = (a-b)/2i$. Since $x(M) \in \mathbb{R}$ and $y(M) \in \mathbb{R}$ for all $M \in \mathcal{M}$, the spectra $\sigma(x, \hat{\mathcal{K}}_0)$ and $\sigma(y, \hat{\mathcal{K}}_0)$ of the elements x and y in the algebra $\hat{\mathcal{K}}_0$ are real. Moreover, $xy = yx$. It follows from Lemma 13.3 that $x + iy \in G\hat{\mathcal{K}}_0$ if and only if $x - iy \in G\hat{\mathcal{K}}_0$. If $A \in \Phi(E) \cap \mathcal{K}$, then

$$\hat{A} \in G\widehat{\mathcal{L}(E)} \cap \hat{\mathcal{K}} \Longrightarrow \mu(\hat{A}) \in G\hat{\mathcal{L}}_0^{n \times n} \cap \hat{\mathcal{K}}_0^{n \times n} \Longrightarrow det\ \mu(\hat{A}) \in G\hat{\mathcal{L}}_0 \Longrightarrow x + iy \in G\hat{\mathcal{L}}_0.$$

Hence, by the preceding discussion, $b = x - iy \in G\hat{\mathcal{L}}_0$. Since $b = det\ \mu(\hat{B})$, Theorem 1.1 implies $\mu(\hat{B}) \in G\hat{\mathcal{L}}_0^{n \times n}$. Finally, we observe that

$$det\ \nu_M(\hat{A}\hat{B} - \lambda e) = det\ (\nu_M(\hat{A})\nu_M(\hat{A})^* - \lambda \nu_M(e)) \neq 0,$$

for every complex $\lambda \in \mathbb{C}$ with $Im\ \lambda \neq 0$, whence $\sigma(\hat{A}\hat{B}, \hat{\mathcal{L}}) \subset \mathbb{R}$. Therefore, $\sigma(\hat{A}\hat{B}, \hat{\mathcal{L}}) = \sigma(\hat{A}\hat{B}, \hat{\mathcal{K}})$, and hence $\hat{A}\hat{B} \in G\hat{\mathcal{K}}$. Consequently, there is an $X \in \mathcal{K}$ such that $\hat{A}\hat{B}\hat{X} = \hat{I}$, whence $\hat{A}^{-1} = \hat{B}\hat{X} \in \hat{\mathcal{K}}$, as claimed.

3) Let $p_1, \ldots, p_n$ and v be elements of the algebra $\hat{\mathcal{K}}$ whose existence is guaranteed by the assumption that $\hat{\mathcal{K}} \in \omega_n$ (see Sec.23), and let $P_1, \ldots, P_n, V$ be operators such that $P_j \in p_j$, $j = 1, \ldots, n$, and $V \in v$. Then equality (23.5) can be reexpressed as

$$[\varepsilon^{(m-1)(k-1)} V_1^{k-1}][A_{jk}][\varepsilon^{(1-m)(k-1)} V^{m-1}] = n\ diag\ (A_1, \ldots, A_n) + T \qquad (26.5)$$

where $T \in \mathcal{T}(E)$, V_1 is a regularizer for V, and

$$A_m = \left(\sum_j \varepsilon^{j(m-1)} P_j \right) A \left(\sum_j \varepsilon^{j(1-m)} P_j \right). \qquad (26.6)$$

The outer factors in the right-hand side of (26.6) have the property that their n-th powers differ from the identity by compact operators. Consequently, $Ind\ A_m = Ind\ A$. Since the entries of the matrix $[\varepsilon^{(m-1)(k-1)} V_1^{(k-1)}]$ pairwise commute, it follows (by Theorem 3.5) that

$$Ind[\varepsilon^{(m-1)(k-1)} V_1^{k-1}] = Ind\ det[\varepsilon^{(m-1)(k-1)} V_1^{(k-1)}]$$
$$= Ind\ V_1^{\frac{n(n-1)}{2}} = \frac{n(n-1)}{2} Ind\ V_1 .$$

Similarly,

$$Ind\ [\varepsilon^{(1-m)(k-1)}V^{m-1}] = \frac{n(n-1)}{2}\ Ind\ V.$$

The needed equality $Ind[A_{jk}] = n\ Ind\ A$ now follows from (26.5) and the fact that $Ind\ V + Ind\ V_1 = 0$.

4) It remains to show that if $\Phi(E) \cap \mathcal{K}$ is dense in $\mathcal{K}$, then $Ind[A_{jk}] = Ind\ det[A_{jk}]$. But this was proved in Sec.3 (see Theorem 3.1).

□

We next give two criteria for the existence of an n-symbol in C^*-algebras of operators.

THEOREM 26.3. *Let E be a Hilbert space and $\mathcal{K}$ a self-adjoint subalgebra of $\mathcal{L}(E)$. Then $\mathcal{K}$ is an algebra with n-symbol (for some value of n) if and only if $\hat{\mathcal{K}} = \mathcal{K}/T(E)$ is a PI-algebra.*

PROOF. The sufficiency of this condition follows from Theorems 22.2 and 26.1. Let us prove its necessity. Let $\{\gamma_M\}$ be a family of homomorphisms which defines a symbol on $\mathcal{K}$. We claim that $\gamma_M(T) = 0$ for all $T \in T(E)$ and all M. Suppose $\gamma_M(T) \neq 0$. Then there is a point $\lambda \neq 0$ in the spectrum of the matrix $\gamma_M(T^*T)$ such that $det\ \gamma_M(\lambda I - T^*T) = 0$, which is impossible because $\lambda I - T^*T \in \Phi(E)$. Thus, the formula $\nu_M(\hat{A}) = \gamma_M(A)$, $A \in \hat{A}$, correctly defines a sufficient family of n-dimensional representations of the quotient algebra $\hat{\mathcal{K}}$. Since $\hat{\mathcal{K}}$ is also semisimple, Theorem 22.2 shows that it is a PI-algebra.

□

In much the same way one can use Theorem 22.1 to prove the following result:

THEOREM 26.4. *Let E be a Hilbert space and $\mathcal{K}$ a self-adjoint subalgebra of $\mathcal{L}(E)$. Then $\mathcal{K}$ is an algebra with n-symbol if and only if the operator $T = F_{2n}(A_1,\ldots,A_{2n})$ is compact for every choice of $A_1,\ldots,A_{2n} \in \mathcal{K}$ (here F_{2n} is the standard polynomial (20.1)).*

□

For operators acting in Banach spaces we have

THEOREM 26.5. *Let $\mathcal{K} \subset \mathcal{L}(E)$ and $\mathcal{K} \in \sigma_n$. Then for arbitrary $A_1,\ldots,A_{2n} \in$*

$\mathcal{K}$ the operator $T = F_{2n}(A_1, \ldots, A_{2n})$ is a Φ-admissible perturbation for the algebra $\mathcal{K}$. Moreover, T is not necessarily compact, but it is compact if $\hat{\mathcal{K}}$ is inverse-closed in $\widehat{\mathcal{L}(E)}$.

PROOF. The proof is identical to that for the case $n = 1$ (Theorem 14.3).

$\square$

Remark 26.1. In Sec.14 we have shown that the condition $[A_1, A_2] = A_1 A_2 - A_2 A_1 \in \mathcal{T}(E)$ for all $A_1, A_2 \in \mathcal{K}$ is necessary and sufficient for $\mathcal{K}$ to belong to the class σ_1. It turns out that for $n \geq 2$ the condition $F_{2n}(A_1, \ldots, A_{2n}) \in \mathcal{T}(E)$ for all $A_1, \ldots, A_{2n} \in \mathcal{K}$ does not necessarily guarantee that $\mathcal{K} \in \sigma_n$. To see this, consider the algebra $\mathcal{K}$ constructed in Example 14.2. It is readily checked that for arbitrary $A_1, \ldots, A_4 \in \mathcal{K}$

$$[A_1, A_2][A_3, A_4] \in \mathcal{T}(E). \tag{26.7}$$

By equality (20.2), this implies that $F_{2n}(A_1, \ldots, A_{2n}) \in \mathcal{T}(E)$ for all $n \geq 2$. We show that $\mathcal{K} \notin \sigma_n$ for any value of n. Suppose, on the contrary, that $\mathcal{K} \in \sigma_m$ for some m, and let $\{\gamma_M\}$ be the corresponding family of homomorphisms. Set $\tilde{\gamma}_M = h_M \gamma_M$, where $h_M \colon Im\ \gamma_M \rightarrow Im\ \gamma_M / \mathcal{R}(Im\ \gamma_M)$ is the natural homomorphism. The family $\{\tilde{\gamma}_M\}$ defines a symbol on $\mathcal{K}$, because $det\ \gamma_M(x) \neq 0$ if and only if $det\ \tilde{\gamma}_M(x) \neq 0$ ($det\ \tilde{\gamma}_M(x)$ is meaningful since the quotient algebra $Im\ \gamma_M / \mathcal{R}(Im\ \gamma_M)$ of the finite-dimensional algebra $Im\ \gamma_M$ can be identified with a semisimple subalgebra of $Im\ \gamma_M$). It follows from (26.7) that $[A_1, A_2]^2 \in \mathcal{T}(E)$ for all $A_1, A_2 \in \mathcal{K}$, which in turn implies that $I - C[A_1, A_2]^2 \in \Phi(E)$ for all $A_1, A_2, C \in \mathcal{K}$. Therefore, $e - \tilde{\gamma}_M(C)[\tilde{\gamma}_M(A_1), \gamma_M(\tilde{A}_1)]^2 \in G\ Im\ \tilde{\gamma}_M$, whence $[\tilde{\gamma}_M(A_1), \tilde{\gamma}_M(A_2)]^2 \in \mathcal{R}(Im\ \tilde{\gamma}_M)$. Since the matrix algebra $Im\ \tilde{\gamma}_M$ is semisimple, $[x, y]^2 = 0$ for all $x, y \in Im\ \tilde{\gamma}_M$. But the semisimple matrix algebra $Im\ \tilde{\gamma}_M$ is a direct sum of simple subalgebras, i.e., of matrix algebras $\mathbb{C}^{m \times m}$. Each of these satisfies the identity $[x, y]^2 = 0$, which is possible only for $m = 1$ (it is readily verified ⟍ that the identity is not satisfied for the pair of matrix units e_{12}, e_{21}). Correspondingly, $\tilde{\gamma}_M$ decomposes into a direct sum of one-dimensional representations. We have thus shown that if $\mathcal{K} \in \sigma_m$ for some m, then $\mathcal{K} \in \sigma_1$. But in Example 14.2 it was shown that $\mathcal{K} \notin \sigma_1$.

Remark 26.2. In Sec.14 it was shown that every commutative subalgebra $\hat{\mathcal{K}}$ of $\widehat{\mathcal{L}(E)}$ can be imbedded in a commutative inverse-closed subalgebra of $\widehat{\mathcal{L}(E)}$. Using this observation we have concluded that if $\hat{\mathcal{K}}$ is commutative then $\mathcal{K} \in \sigma_1$. The example that we have just constructed shows that not every PI-subalgebra of $\widehat{\mathcal{L}(E)}$ can be imbedded in an inverse-closed PI-subalgebra of $\widehat{\mathcal{L}(E)}$.

We conclude this section by deriving a condition under which the order of the symbol cannot be reduced. To this end we need the following definition:

Definition [67, p. 4]. The polynomial $f(x_1, \ldots, x_r)$ is said to be a *central polynomial* on the algebra $\mathcal{K}$ if $f(a_1, \ldots, a_r)$ belongs to the center of $\mathcal{K}$ for all $a_1, \ldots, a_r \in \mathcal{K}$ and there exist $a_1^0, \ldots, a_r^0 \in \mathcal{K}$ such that $f(a_1^0, \ldots, a_r^0) \neq 0$. For $\mathbb{C}^{n \times n}$ the center coincides with the set of scalar multiples of the identity matrix. It is readily verified that the polynomial $f(x, y) = [x, y]^2$ is central on $\mathbb{C}^{2 \times 2}$ (this has already been remarked above). It is interesting that until 1972 no central polynomial was known for the algebras $\mathbb{C}^{n \times n}$ with $n > 2$. The existence of such polynomials was first established by Formanek [16] and Razmyslov [45 (8)]. For further investigations in this direction see [67, p. 24].

THEOREM 26.6. *Let $\mathcal{K} \in \sigma_n$ and let $\{\gamma_M\}$ be a family of homomorphisms which defines an n-symbol on $\mathcal{K}$. Then $\mathcal{K} \notin \sigma_m$ for all $m < n$ if and only if there is an M_0 such that $Im\ \gamma_{M_0} = \mathbb{C}^{n \times n}$.*

PROOF. Sufficiency. Let $\varphi(x_1, \ldots, x_r)$ be a central polynomial on $\mathbb{C}^{n \times n}$ and set $A = I - \varphi(A_1, \ldots, A_r)$, where $A_1, \ldots, A_r \in \mathcal{K}$ are chosen so that $\varphi(\gamma_{M_0}(A_1), \ldots, \gamma_{M_0}(A_r)) = e$. Suppose that $\mathcal{K} \in \sigma_m$ for some $m < n$, and let $\{\xi_s\}$ be the corresponding family of homomorphisms. It is easily seen that $\varphi(z_1, \ldots, z_r) = 0$ for every collection of matrices $z_1, \ldots, z_r \in \mathbb{C}^{m \times m}$. Then $det\ \xi_s(A) = 1$ for all ξ_s, and hence $A \in \Phi(E)$. But $\gamma_{M_0}(A) = 0$, and hence $A \notin \Phi(E)$. This proves the sufficiency of the indicated condition.

The necessity of this condition is established as in Remark 26.1: every representation γ_M for which $Im\ \gamma_M = \mathbb{C}^{n \times n}$ can be replaced by a number of irreducible representations of dimensionalities less than n.

$\square$

27. MATRIX SYMBOLS ON ALGEBRAS THAT CONTAIN NON-CARLEMAN SHIFTS

Example 27.1. Let Γ be a simple closed contour in the complex plane, β a homeomorphism of Γ onto itself, $\beta_{-1} = \beta^{-1}$, and suppose that $\beta', \beta'_{-1} \in L_\infty(\Gamma)$. Then the shift operator $H \colon (H\varphi)(t) = \varphi(\beta(t))$ is bounded and invertible in $E = L_p(\Gamma)$ $(1 \le p \le \infty)$. As we have already mentioned, if $H^n = I$ for some integer $n > 0$, then β is referred to as a *Carleman shift*. Let B denote the operator of multiplication by the independent variable: $(B\varphi)(t) = t\varphi(t)$. Numerous examples of algebras are known which contain the operators B and H, and on which one can introduce a matrix symbol (see, e.g., [53, Ch. VIII]). In all these examples β is a Carleman shift.

Let $\mathcal{K}$ be a subalgebra of $\mathcal{L}(E)$ such that $B, H \in \mathcal{K}$.

THEOREM 27.1. *If $H^k \ne I$ for all positive integers k, then on algebra $\mathcal{K}$ one cannot introduce an n-symbol for any n.*

PROOF. Suppose that for an integer $n > 0$ there is an n-symbol on $\mathcal{K}$. Set $\beta_1 = \beta$, $\beta_k(t) = \beta(\beta_{k-1}(t))$, $k = 2, 3, \dots$. Since β is a non-Carleman shift, there exist a $t_0 \in \Gamma$ such that the points $t_0, \beta_1(t_0), \dots, \beta_\ell(t_0)$ are all distinct (where $\ell = 2n(4n^2 - 1)/3$). Let f be a polynomial with the property that $f(\beta_k(t_0)) = exp(i\ell k)$. Consider the operator $A_\lambda = \lambda H^m - F_{2n}(fH, fH^2, \dots, fH^{2n})$, where $m = n(1 + 2n)$, $\lambda \ne 0$, and F_{2n} is the standard polynomial (20.1).

Let $\{\gamma_M\}$ be a family of homomorphisms which defines a symbol on $\mathcal{K}$. Then $\gamma_M(A_\lambda) = \lambda \gamma_M(H^m)$. Since $H^m \in GL(E)$, $det\ \gamma_M(H^m) \ne 0$ for all M, and hence $det\ \gamma_M(A_\lambda) \ne 0$ for all M, which in turn implies that A_λ is a Fredholm operator for all $\lambda \ne 0$. It is readily checked that

$$F_{2n}(fH, fH^2, \dots, fH^{2n}) = \sum \pm fH^{p_1} fH^{p_2} \dots fH^{p_n}$$

$$= \sum \pm f f_{p_1 + p_2} \dots f_{p_1 + \dots + p_{2n-1}} H^m = hH^m,$$

where the sums are taken over all permutations $p_1, \dots, p_{2n}$ of the numbers $1, \dots, n$ and $f_k(t) = f(\beta_k(t))$.

Therefore, $A_\lambda = (\lambda - h)H^m$. It follows that the operator of multiplication by the function $\lambda - h$ is Fredholm for every $\lambda \neq 0$. This is possible only when $h(t) \equiv 0$. But

$$h(t_0) = \sum \pm exp\{i(p_{2n-1} + 2p_{2n-2} + \ldots + (2n-1)p_1)\}$$
$$= -exp(i\ell) + \sum_{k<\ell} b_k exp(ik),$$

where $b_k \in \mathbb{Z}$ and $\ell = \sum_{k=2}^{2n}(k-1)k = 2n(4n^2 - 1)/3$ and hence, since $exp\ i$ is a transcendental number, $h(t_0) \neq 0$, which is a contradiction.

□

Remark 27.1. Theorem 27.1 remains valid in a more general situation. Specifically, $L_p(\Gamma)$ may be replaced by an arbitrary Banach space of functions defined on a topological space Γ. However, the following requirement must be fulfilled: **a)** the algebra $\mathcal{K}$ contains a non-Carleman shift operator based on a homeomorphism β of Γ, and also operators of multiplication by continuous functions which separate the points $t_0, \beta(t_0), \ldots, \beta_m(t_0)$, and **b)** if $A = aI \in \mathcal{K}$ is a multiplication operator and $A \in \Phi(E)$, then $a(t) \neq 0 \quad \forall\, t \in \Gamma$.

Example 27.2. Consider again the algebra $\mathcal{K}_3$, defined in Sec.24, which is generated by discrete multiplication operators and a non-Carleman shift. It is not hard to show that the quotient algebra $\mathcal{K}_3/T(E)$ is commutative, and hence $\mathcal{K}_3 \in \sigma_1$. Theorem 27.1 fails for $\mathcal{K}_3$ because this algebra contains Fredholm operators of multiplication by sequences with finitely many nonzero elements.

Example 27.3. We give an example of an algebra $\mathcal{K}$ which contains a non-Carleman shift operator as well as multiplication operators, such that $\mathcal{K}$ does not admit an n-symbol for any value of n, but admits a matrix symbol with no constraints on the orders of the representations.

Let Γ be a compactum. We call the homeomorphism $\beta\colon \Gamma \to \Gamma$ a *locally Carleman homeomorphism* if for every point t_0 the orbit $t_0,\ t_1 = \beta(t_0), t_2 = \beta(t_1), \ldots$ is finite.

As an example we consider the compactum $\Gamma = \bigcup_{n\in\mathbb{N}\cup\{\infty\}} \Gamma_n$, where $\Gamma_n = \{\zeta \in \mathbb{C}\colon |\zeta| = 1 + 1/n\}$, and the homeomorphism β defined by the formula $\beta(t) = $

$t \, exp \, (\pi i/n)$ for $t \in \Gamma_n$. The orbit of any point $t \in \Gamma_n$ ($n \in \mathbb{N}$) consists of $2n$ points, and the circle Γ_∞ is fixed by β. Hence, β is a locally Carleman shift on Γ. Consider the algebra $\mathcal{K} \subset \mathcal{L}(L_2(\Gamma))$ generated by the operators of multiplication by functions $f \in C(\Gamma)$ and by the shift operators W and W^{-1}, where $(W\varphi)(t) = \varphi(\beta(t))$. In $\mathcal{K}$ every Fredholm operator is invertible [52]. Moreover, $\mathcal{K}$ is an inverse-closed subalgebra in $\mathcal{L}(E)$ (where $E = L_2(\Gamma)$). This implies that a sufficient family of finite-dimensional representations defines a matrix symbol on $\mathcal{K}$. Let $\mathcal{K}_0$ denote the subalgebra of $\mathcal{K}$ consisting of the operators of multiplication by functions which are constant on each circle Γ_n, $n \in \mathbb{N} \cup \{\infty\}$. It is readily verified that $\mathcal{K}_0$ is a subalgebra of the center of $\mathcal{K}$. The set M_n, where $n \in \mathbb{N} \cup \{\infty\}$, of operators (from $\mathcal{K}_0$) of multiplication by functions which are equal to zero on Γ_n is a maximal ideal in $\mathcal{K}_0$. One can show that these are the only maximal ideals of $\mathcal{K}_0$. Following the scheme proposed in Sec.24, we consider the quotient algebras $\mathcal{K}/J_{M_n} = \mathcal{K}_n$. It is easily seen that for $n \in \mathbb{N}$ the algebra $\mathcal{K}_n$ is isomorphic to the algebra $\overset{\circ}{\mathcal{K}}_n \subset \mathcal{L}(L_2(\Gamma_n))$ generated by the operators of multiplication by functions $f \in C(\Gamma_n)$ and by the operator of rotation by the angle π/n. It is known that $\overset{\circ}{\mathcal{K}}_n \in \pi_{2n}$ (see [53]). Consequently, $\mathcal{K}_n \in \pi_{2n}$ for every $n \in \mathbb{N}$. $\mathcal{K}_\infty$ is isomorphic to the commutative algebra $\overset{\circ}{\mathcal{K}}_\infty$ generated by the operators W and W^{-1} in $L_2(\Gamma)$, and hence $\mathcal{K}_\infty \in \pi_1$. By Theorem 24.1, if $A \in \mathcal{K}$, then $A \in G\mathcal{K}$ if and only if $A_n \in G\mathcal{K}_n$ for all n (where $A \in A_n \in \mathcal{K}_n$). Each homomorphism $\gamma_{M_n} : \mathcal{K}_n \to \mathbb{C}^{\ell(M) \times \ell(M)}$ naturally induces a **homomorphism** $\tilde{\gamma}_{M_n} : \mathcal{K} \to \mathbb{C}^{\ell(M) \times \ell(M)}$, and the family $\{\tilde{\gamma}_{M_n}\}_{n \in \mathbb{N} \cup \{\infty\}}$ defines a symbol on $\mathcal{K}$.

28. NONEXISTENCE OF A MATRIX SYMBOL ON THE ALGEBRA OF MULTIDIMENSIONAL WIENER-HOPF OPERATORS

It was shown in Sec.14 that the algebra $\mathcal{K}$ generated by the multidimensional Wiener-Hopf operators does not admit a scalar symbol. Here we show that this algebra does not admit a matrix symbol either, for any $n \in \mathbb{N}$.

Let $E = L_p(\mathbb{R}_+^{m+1})$, where $1 \leq p \leq \infty$ and $m \in \mathbb{N}$.

THEOREM 28.1. *Let $\mathcal{K}_p \subset \mathcal{L}(E)$ be an arbitrary algebra which contains all Wiener-Hopf operators A:*

$$(A\varphi)(t) = \varphi(t) + \int_{\mathbb{R}_+^{m+1}} k(t-s)\varphi(s)ds, \qquad t \in \mathbb{R}_+^{m+1}, \tag{28.1}$$

where $k \in L_1(\mathbb{R}^{m+1})$. Then $\mathcal{K}$ does not admit an n-symbol.

We need the following lemma:

LEMMA 28.1. *Let $V \in \mathcal{L}(E)$ be a left-invertible operator, let $V^{(-1)}$ be a left inverse of V, $e_1 \in Ker\ V^{(-1)}$ $(e_1 \neq 0)$, and $e_k = V^{k-1}e_1$ and*

$$A_k = V^k - V^{(-k)}. \tag{28.2}$$

If $0 < p_1 < p_2 < \ldots < p_n$, then

$$F_n(A_{p_1}, A_{p_2}, \ldots, A_{p_n})e_1 = (-1)^{n(n-1)/2}e_r, \tag{28.3}$$

where $r = 1 + \sum_{k=1}^{n}(-1)^{n-k}p_k$.

PROOF. We proceed by induction. For $n = 1$ (28.3) is obvious. Using equality (20.1), we get, by the inductive hypothesis,

$$\begin{aligned}
F_n(A_{p_1}, A_{p_2}, \ldots, A_{p_n})e_1 &= \sum_{j=1}^{n}(-1)^{j-1}(-1)^{(n-1)(n-2)/2}A_{p_j}e_{q_j} \\
&= (-1)^{(n-1)(n-2)/2}\sum_{j=1}^{n}(-1)^{j-1}(e_{q_j+p_j} - e_{q_j-p_j})
\end{aligned} \tag{28.4}$$

where

$$q_j = 1 + \sum_{k=1}^{j-1}(-1)^{n-k-1}p_k + \sum_{k=j}^{n-1}(-1)^{n-k-1}p_{k+j}, \qquad j = 1, 2, \ldots, n,$$

and for $m \leq 0$ we put $e_m = 0$. Now we set

$$r_j = 1 + \sum_{k=1}^{j}(-1)^{n-k-1}p_k + \sum_{k=j+1}^{n}(-1)^{n-k}p_k, \qquad j = 0, 1, \ldots, n,$$

and notice that

$$r_0(=r) = 1 + \sum_{k=1}^{n}(-1)^{n-k}p_k$$

and

$$r_n = 1 + \sum_{k=1}^{n}(-1)^{n-k-1}p_k < 0$$

$$(28.5)$$

It is readily checked that the pairs of numbers (q_j+p_j, q_j-p_j), $j = 1,\ldots,n$, coincide with the pairs (r_{j-1}, r_j) for $n-j$ even, and (r_j, r_{j-1}) for $n-j$ odd. Consequently,

$$e_{q_j+p_j} - e_{q_j-p_j} = (-1)^{n-j}(e_{r_{j-1}} - e_{r_j}), \qquad j = 1, 2, \ldots, n.$$

Hence, equality (28.4) yields

$$f_n(A_{p_1}, A_{p_2}, \ldots, A_{p_n})e_1 = (-1)^{n(n-1)/2} \sum_{j=1}^{n}(e_{r_{j-1}} - e_{r_j}),$$

whence, in view of (28.5), $F_n(A_{p_1}, A_{p_2}, \ldots, A_{p_n})e_1 = (-1)^{n(n-1)/2}e_r$, as claimed.

□

COROLLARY 28.1. *If the assumptions of Lemma* 28.1 *are in force, then*

$$F_{2n}(A_1, A_2, \ldots, A_{2n})e_1 = (-1)^n e_{n+1}.$$

$$(28.6)$$

□

PROOF OF THEOREM 28.1. Consider in $L_p(0,\infty)$ the Wiener-Hopf operators

$$(Vf)(x) \quad = f(x) - 2\int_0^x e^{y-x}f(y)dy, \qquad 0 \le x < \infty,$$

$$(V^{(-1)}f)(x) = f(x) - 2\int_x^\infty e^{x-y}f(y)dy, \qquad 0 \le x < \infty,$$

and

$$A_r = V^r - V^{(-r)}, \qquad r = 1, 2, \ldots \; .$$

Also, let $a_r \in L_1(\mathbb{R})$ denote the function defined by the equality

$$(A_r f)(x) = \int_0^\infty a_r(x - y)f(y)dy ,$$

and set

$$(B_r \varphi)(x) = \int_{\mathbb{R}_+^{m+1}} k_r(t - s)\varphi(s)ds, \qquad t \in \mathbb{R}_+^{m+1},$$

for $r = 1, 2, \ldots$, where

$$k_r(t) = a_r(x)exp\left(-\sum_{j=1}^m |\tau_j|\right),$$

$$t = (x, \tau), \qquad x \in \mathbb{R}, \qquad \tau = (\tau_1, \ldots, \tau_m) \in \mathbb{R}^m.$$

If the function φ has the form $\varphi = fg$, with $f \in L_p(\mathbb{R}_+)$ and $g \in L_p(\mathbb{R}^m)$, then

$$(B_r \varphi)(t) = (A_r f)(x)(C_g)(\tau)$$

where C is the operator in $L_p(\mathbb{R}^m)$ defined by the formula

$$(Cg)(\tau) = \int_{\mathbb{R}^m} exp\left(-\sum_{j=1}^m |\tau_j - \mu_j|\right)g(\mu)d\mu, \qquad \tau \in \mathbb{R}^m.$$

It is readily verified that

$$(F_r(B_1, \ldots, B_r)\varphi)(t) = (F_r(A_1, \ldots, A_r)f)(x)(C^r g)(\tau). \qquad (28.7)$$

Lemma 28.1 applies to the operators V, $V^{(-1)}$ and the vector $e_1 = e^{-x} \in Ker\ V^{(-1)}$ and upon setting $f(x) = e^{-x}$ and $g(\tau) = exp(-\sum |\tau_j|)$ in (28.7) yields $F_r(B_1, \ldots, B_r) \neq 0$, $r = 1, 2, \ldots$.

We consider first the case $p = 2$ and assume that the algebra $\mathcal{K}_2$ generated by the operators of the form (28.1) admits an n-symbol for a positive integer n.

Since the operator $R = F_{2n}(B_1, \ldots, B_{2n}) \neq 0$, there exists a number $\lambda_0 > 0$ such that $A = \lambda_0 I - R^* R$ is not invertible. The operator A can be expressed in the form $\lambda_0 I - \sum B_{j1} \ldots B_{jr}$, where B_{jk} are Wiener-Hopf operators. Its linear dilation $\Xi(A)$ (see

Sec.2) is a multidimensional Wiener-Hopf operator with matrix kernel. In [25(6)] it is proved that if such an operator is a Φ_+- or Φ_--operator, then it is not invertible. This shows, in particular, that $A = \lambda_0 I - R^* R \notin \Phi(E)$. Let $\mathcal{K}_2 \in \sigma_n$. Then, by Theorem 26.4, $F_{2n}(B_1, \ldots, B_{2n}) \in \mathcal{T}(E)$, which contradicts the fact that $A \notin \Phi(E)$. This completes the proof of the theorem for $p = 2$.

Now let $p < 2$. The operators R and R^* belong to $\mathcal{L}(L_r(\mathbb{R}_+^{m+1}))$ for every r ($1 \leq r \leq \infty$). If the algebra $\mathcal{K}_p$ admits an n-symbol, then the operator $A = \lambda_0 I - R^* R$ defined above is Fredholm in $L_p(\mathbb{R}_+^{m+1})$ (because $det \gamma_M(A) = \lambda_0^n \neq 0 \; \forall M$). Since $A^* = A$, we see that $A \in \Phi(L_q(\mathbb{R}_+^{m+1}))$ and $Ind\, A \mid L_p = Ind\, A \mid L_q = 0$. It is readily verified that $F_{2n}(B_1, \ldots, B_{2n})^* F_{2n}(B_1, \ldots, B_{2n})(L_p) \subset L_p \cap L_\infty \subset L_r$ for all $r \in (p, \infty)$. Consequently, $Ker\, A \mid L_p = Ker\, A \mid L_q$. Let $\varphi_k \in L_p \cap L_q$, $k = 1, \ldots$, be an L_2-orthonormal basis in $Ker\, A$ and set

$$Cx = Ax - \sum_{j=1}^{\ell} (x, \varphi_j) \varphi_j.$$

It is easily seen that the operator C is invertible in both L_p and L_q ($Ind\, C = 0$ and $dim\, Ker\, C = 0$), and hence in L_2. Therefore $A \in \Phi(L_2)$, which, as we already observed, implies that A is invertible in L_2, which is false.

The case $p > 2$ is dealt with by passing to the dual space.

$\square$

Remark 28.1. Establishing the fact that the algebra generated by discrete multidimensional Wiener-Hopf operators does not admit a matrix symbol is considerably simpler. The reason is that among such operators there are operators invertible only from one side with infinite-dimensional defect space (for example, the shift operator [51]). Therefore, one can use the following

Remark 28.2. Suppose that the algebra $\mathcal{K} \subset \mathcal{L}(E)$ contains operators A and B such that $AB = I$ and $dim\, Ker\, A = \infty$. Then $\mathcal{K}$ does not admit an n-symbol for any n.

In fact, in this case we would have $\gamma_M(A)\gamma_M(B) = \gamma_M(I)$ and hence $det\, \gamma_M(A) \neq 0 \;\; \forall M$, but $A \notin \Phi(E)$.

In the algebra $\mathcal{K}_p$ generated by continual multidimensional Wiener-Hopf operators there are no operators invertible only from one side, so that Remark 28.2 is inapplicable.

We conclude with two examples of operator algebras in which Remark 28.2 can be used.

Example 28.1. Let V denote the operator in $L_p(0, \infty)$, with $1 \le p \le \infty$, defined by the formula

$$(Vf)(t) = \begin{cases} f(t-1) & \text{if} \quad t > 1, \\ 0 & \text{if} \quad t < 1. \end{cases}$$

The operator $(V^{(-1)}f)(t) = f(t+1)$, $t \in (0, \infty)$, is a left inverse of V. Since $\dim \operatorname{Ker} V^{(-1)} = \infty$, no algebra which contains V and $V^{(-1)}$ admits an n-symbol for any value of n (Remark 28.2). This is true, in particular, for the algebra generated by the finite-difference operators of the form

$$(Af)(t) = \sum_{j=-\infty}^{\infty} a_j f(t-j) \,,$$

where $\sum_{j=-\infty}^{\infty} |a_j| < \infty$ and $f(t-j) = 0$ for $0 \le t \le j$.

Example 28.2. Let Γ_0 denote the unit circle. In the Banach space $E = L_p(\Gamma_0)$ with $1 < p < \infty$, consider the operators $A = aP + Q$ and $B = \bar{a}P + Q$, where $P = \frac{1}{2}(I+S)$, $Q = \frac{1}{2}(I-S)$, S is the operator of singular integration along Γ_0, and a is a singular inner function [28]. It is readily verified that $BA = I$. We show that $\dim \operatorname{Ker} B = \infty$. The singular inner function $\bar{a}$ can be uniformly approximated by Blaschke products b_n each of which has infinitely many zeros in the disc $|z| < 1$ (see [28]). The operators $B_n = \bar{b}_n P + Q$ are right invertible $((\bar{b}_n P + Q)(b_n P + Q) = I)$ and $\dim \operatorname{Ker} B_n = \infty$. Since $\|B_n - B\| \to 0$, it follows that $B \notin \Phi(E)$. Hence, by Remark 28.2, no algebra $\mathcal{K}$ which contains the operators A and B admits an n-symbol.

29. OPEN PROBLEMS

We formulate a number of problems related to the theme of this book.

1. Is it possible to replace the ideal of trace class operators by some larger ideal of algebra $\mathcal{L}(H)$ in Theorem 3.4?

2. Let $P = 2S_0 - I$ denote the analytic projector. In Sec. 4 (see (4.4)), it is shown that $|P|_{p,\rho} \geq [sin(\pi/r)]^{-1}$, where $\rho(t) = |t - t_0|^\beta$, $-1 < \beta < p - 1$, $r = \max(p, p(p-1)^{-1}, p(1+\beta), p(p-1-\beta)^{-1})$.

Conjecture. $|P|_{p,\rho} = [sin(\pi/r)]^{-1}$.

This conjecture has been proved only for $p = 2$ and $-1 < \beta < 1$ (Theorem 4.2).

A more general conjecture (also confirmed only for $p = 2$) is: *Estimates (4.2) and (4.4) are exact.*

3. It would be interesting to obtain analogs of the results of Sec.12 for singular integral operators with operator or matrix coefficients.

4. In connection with Problem 3 we formulate the following:

Conjecture. Let $A \in L_\infty(\mathcal{L}(H))$. Then the operator $AP + Q$ is invertible in the spaces $L_p(\Gamma_0, H)$ and $L_q(\Gamma_0, H)$ (with $p^{-1} + q^{-1} = 1$) if and only if $A = B_1^* U B_2$, where $B_k \in GL_\infty^+(\mathcal{L}(H))$, $k = 1, 2$, U is unitary-valued, and there is an operator-valued function $B \in GL_\infty^+(\mathcal{L}(H))$ such that

$$\| U - B \|_{L_\infty(\mathcal{L}(H))} < sin\frac{\pi}{\max(p, q)}.$$

The validity of this conjecture for $p = 2$ was established in [65], and in Theorem 12.1 for $H = \mathbb{C}$ and any $p \in (1, \infty)$. The sufficiency of the indicated conditions for arbitrary H and all $p \in (0, 1)$ follows from Theorem 11.2.

5. The criterion for the invertibility of the operator $h^{-1}S_0 h$ in $L_p(\Gamma_0)$ obtained in [29] plays an important role in problems pertaining to the invertibility of operators of the form $aI + bS_0$ (see Theorem 9.4). It would be interesting to obtain a criterion for the invertibility of operators of the form $A_+ S_0 A_+^{-1}$ in $L_p(\Gamma_0, H)$, where $A_+^{\pm 1} \in L_1^+(\mathcal{L}(H))$.

As far as we know, this problem has not been solved even for *dim H* $< \infty$.

6. Let E $(\subset L_1(\Gamma_0))$ be a Banach space in which the operator S_0 and the operator R_Γ of multiplication by the indicator function χ_Γ of every arc $\Gamma \subset \Gamma_0$ are bounded.

Let $\rho \in E$, $\rho^{-1} \in E^*$, and let $E(\rho)$ denote the Banach space of functions φ with the norm $\| \varphi \|_{E(\rho)} = \| \rho\varphi \|_E$.

Conjecture. The operator S_0 is bounded in $E(\rho)$ if and only if there is a number $c = c(E)$ such that $\| \chi_\Gamma \rho \|_E \| \chi_\Gamma \rho^{-1} \|_{E^*} \leq c$ mes Γ for all arcs $\Gamma \subset \Gamma_0$.

This conjecture arose in connection with the work [29], in which the assertion was proved for the spaces $E = L_p(\Gamma_0)$ with $1 < p < \infty$.

7. Let Γ be a simple closed Lyapunov contour. Then there exists a homeomorphism $\alpha\colon \Gamma_0 \to \Gamma$ such that, if $a, b \in L_\infty(\Gamma)$, then $(aI + bS_\Gamma \in \Phi(L_p(\Gamma))) \Longleftrightarrow (a_0 I + b_0 S_0 \in \Phi(L_p(\Gamma_0)))$ where $a_0(z) = a(\alpha(z))$, $b_0(z) = b(\alpha(z))$ (see [50(9), Ch.VII, §2]).

Question: Which non-Lyapunov curves enjoy an analogous property?

8. The algebra generated by one-dimensional singular integral operators, the operators of which have discontinuities of almost-periodic type, does not admit a matrix symbol, because it contains operators invertible from only one side with infinite-dimensional defect space. The criterion for an operator in this algebra to be Fredholm is not known. Such a criterion is known only for the generators [50(9), Ch. XI] or for a product of several generators whose coefficients have no common points of discontinuity [50(9)].

9. Let $\mathcal{K}_0$ be a dense subalgebra of the Banach algebra $\mathcal{K}$ such that $\mathcal{K}_0$ has a sufficient family of n-dimensional representations. Does this imply that $\mathcal{K} \in \pi_n$? Example 18.4 shows that a sufficient family of homomorphisms of $\mathcal{K}_0$ is not necessarily sufficient for $\mathcal{K}$. This, however, does not mean that $\mathcal{K} \notin \pi_n$ (in Example 18.4, $\mathcal{K} \in \pi_1$).

10. Let $\mathcal{K}_0$ be a nonclosed subalgebra of $\mathcal{L}(E)$ which possesses an n-symbol. Is it necessarily true that its closure $\overline{\mathcal{K}}_0 \in \sigma_n$?

11. Under certain restrictions on the algebra $\mathcal{K}_0$ and the family of representations it follows from Theorem 18.5 that the answer to Problem 10 is affirmative. It would be desirable to weaken the assumptions of this theorem, leaving only the essential ones.

12. Theorem 22.1 characterizes the class π_n of Banach algebras.

Problem. Characterize those Banach algebras which possess a sufficient family of finite-dimensional representations with no constraint on the orders of the representations.

13. Every infinite-dimensional Banach algebra admits the trivial "sufficient family of infinite-dimensional representations" which consist of the single **homomorphism** $\gamma: \mathcal{K} \to \mathcal{L}(\mathcal{K})$ defined by the rule $\gamma(a)x = ax$.

Problem. Propose a reasonable definition of a sufficient family of infinite-dimensional representations and characterize the Banach algebras which possess such a family.

14. Relate the solution of the preceding problem to the construction of an operator-valued symbol on operator algebras.

15. Suppose that in algebra $\mathcal{K}_0 \subset \mathcal{L}(E)$ every Fredholm operator is invertible. Does algebra $\mathcal{K}$ enjoy this property? For some results in this direction see [52,52(1)–52(6)].

BRIEF COMMENTS ON THE LITERATURE

Sec.1. Theorem 1.1 was first proved by the author of this book [39] and then extended, in collaboration with S.Sh. Kesler, to the case of one-side invertibility [32]. These results were further developed in works of B. Silberman and U. Köhler (see [53, 64]), A.S. Markus and I.A. Fel'dman [55], and N.L. Vasilevskii and R. Trujilio [75]. In the proof of Theorems 1.2–1.4 we follow [55]. Theorem 1.7 was established in [44(4)].

Sec.2. The results of this section were established simultaneously with those of Sec.1, and in the same works.

Sec.3. Theorem 3.1 was obtained by the author of this book [45], and Theorem 3.4 by S.A. Markus and I.A. Fel'dman [41].

Sec.4. The main results of this section were obtained in [50(1), 50(4), 63, 49, 50, 42, 87, 88]. Theorem 4.4 is taken from Zygmund's book [81, Vol. I].

Sects.5–8,11. The results of these sections were obtained in works of I.E. Verbitskii and the author [77,78], and also in author's works [87], [88].

Sec.10. The criterion for a singular integral operator with piecewise-continuous matrix-valued coefficients was established by I. Gohberg and the author [34(54)].

Sec.12. The results of this section were established in [40,43].

Sec.13. Theorems 13.1–13.3 were formulated by the author in [44,45], and Lemma 13.3 in [39]. Theorems 13.5 and 13.6 are based on results of A.S. Dynin, which appear in [48], and are concerned with the construction of symbols of singular integral operators with Carleman shift along contours with angles.

Sec.14. The results of this section were obtained in [44,47].

Sec.15. The local principle, in the form given in this section, is borrowed from [50(9)].

Sects.16-18. References [53(63-64)] are devoted to the construction of a symbol in algebras generated by Toeplitz and singular integral operators with piecewise-continuous coefficients (see also the literature cited therein). A simple proof of Lemmas 16.2 and 16.3 was proposed by R.V. Duduchava [82], [83].

Sec.19. The results of this section were obtained in [53(66-67)].

Sec.20. Theorems 20.1 and 20.2 were taken from monographs [67] [85], and [27], respectively, which also contain historical comments on the formulation and resolution of problems connected with standard identities.

Sects.21-24,26. The main results of these sections were obtained in [44,45]. The proofs of the auxiliary lemmas in Sec.21 follows the methods of the book [27], where the authors of the various assertions are indicated. Theorem 24.1 for C^*-algebras was obtained by R. Douglas [15].Lemma 24.1 was obtained by G. Allan [3].

Sec.25. The main results of this section are due to M.B. Abalovich and the author [1]. The scheme of the proof of Theorem 25.2 is taken from [14].

Sects.27,28. The nonexistence of a matrix symbol was established in collaboration with I.A. Fel'dman [51].

REFERENCES

[1] M.B. Abalovich and N.Ya. Krupnik: *Topology on the set of maximal ideals of Banach PI-algebras*, in: *Studies on Functional Analysis and Differential Equations, Mathematical Sciences*, "Shtiintsa", Kishinev, 1984, pp. 3-13. (Russian).

[2] M.S. Agranovich: *Elliptic singular integro-differential operators*, Usp. Mat. Nauk **20**, No. 5 (1965), 3-120; English transl.: Russian Math. Surveys **20**, No. 5/6 (1965), 1-121.

[3] G.R. Allan: *One-sided inverses in Banach algebras of holomorphic vector-valued functions*, J. London Math. Soc. **42**, No. 3 (1967), 463-470.

[4] S.A. Amitsur and J. Levitski: *Minimal identities for algebras*, Proc. Amer. Math. Soc., No. 1 (1950), 449-463.

[5] A.B. Antonevich: *On the index of a pseudodifferential operator with a finite group of shifts*, Dokl. Akad. Nauk SSSR **190**, No. 4 (1970), 751-752; English transl.: Soviet Mat. Dokl. **11**, No. 1 (1970), 168-170.

[6] K.I. Babenko: *On conjugate functions*, Dokl. Akad. Nauk SSSR **62**, No. 2 (1948), 157-160. (Russian).

[7] N. Bourbaki: *Éléments de Mathématique, Fasc. 32, Theories Spectrales*, Hermann, Paris, 1967.

[8] A.P. Calderon and A. Zygmund: *Algebras of certain singular operators*, Amer. J. Math. **78**, No. 2 (1956), 310-320.

[9] A.P. Calderon and A. Zygmund: *Singular integral operators and differential equations*, Amer. J. Math. **79**, No. 4 (1957), 901-921.

[10] L.A. Coburn: *The C^*-algebra generated by an isometry*, Bull. Amer. Math. Soc. **73** (1967), 722-726.

[11] R.R. Coifman and C. Fefferman: *Weighted norm inequalities for maximal functions and singular integrals*, Studia Math. **51** (1974), 241-250.

[12] I.I. Danilyuk: *Irregular Boundary Value Problems in the Plane*, "Nauka", Moscow 1975. (Russian).

[13] A. Devinatz: *Toeplitz operators on H_2-space*, Trans. Amer. Math. Soc. **112**, No. 2 (1964), 304-317.

[14] J. Dixmier: *Les C^*-Algebres et leurs Representations*, Gauthier-Villars, Paris, 1964.

[15] R.G. Douglas: *Banach Algebra Techniques in Operator Theory*, Academic Press, New York, 1972.

[16] E. Formanek, *Central polynomials for matrix rings*, J. Algebra **23** (1972), 129-132.

[17] K.O. Friedrichs and P.D. Lax: *Boundary value problems for first order operators*, Commun. Pure Appl. Math. **18**, No. 1/2 (1965), 355-388.

[18] F.D. Gakhov: *Boundary Value Problems*, "Nauka", Moscow, 1977; English transl. of the 2nd edition: Pergamon Press, Oxford, 1966.

[19] F.R. Gantmakher: *Matrix Theory*, "Nauka", Moscow, 1967; English transl. of 1st ed.: Chelsea, New York, 1959.

[20] I.M. Gelfand, D.A. Raikov and G.E. Shilov: *Commutative Normed Rings*, "Fizmatgiz", Moscow, 1959; English transl.: Chelsea, New York, 1964.

[21] G. Giraud: *Equations a integrales principales*, Ann. Scient. Ecole Norm. Super **51**, fasc. 3, 4 (1934), 251-372.

[22] G. Giraud: *Sur certaines operations du type elliptique*, C.R. Acad. Sci. Paris **200**(1935), 1651-1653.

[23] I.M. Glazman and Yu. I. Lyubich: *Finite-Dimensional Linear Analysis*, "Nauka", Moscow (1969); English transl.: The MIT Press, Cambridge, Massachusetts, 1974.

[24] G.H. Hardy and J.E. Littlewood: *Some more theorems concerning Fourier series and Fourier power series*, Duke Math. J. **2** (1936), 354-382.

[25] P.R. Halmos: *A Hilbert Space Book Problem*, 2nd ed., Springer-Verlag, New York, 1982.

[26] H. Helson and G. Szegö: *A problem in prediction theory*, Ann. Mat. Pura Appl. **51** (1960), 107-138.

[27] I.N. Herstein: *Noncommutative Rings*, Carus Math. Monographs **15**, Math. Assoc. of Amer., J. Wiley, 1968.

[28] K. Hoffman: *Banach Spaces of Analytic Functions*, Prentice-Hall, Englewood-Cliffs, N.J., 1962.

[29] R. Hunt, B. Muckenhoupt and R. Wheeden: *Weighted norm inequalities for the conjugate function and Hilbert transform*, Trans. Amer. Math. Soc. **176** (1973), 227-251.

[30] V.V. Ivanov: *Theory of Approximate Methods and its Application to Numerical Resolution of Singular Integral Equations*, "Naukova Dumka", Kiev, 1968. (Russian).

[31] T. Kato: *Perturbation Theory for Linear Operators*, Springer-Verlag, Berlin, 1966.

[32] S. Sh. Kesler and N. Ya. Krupnik: *On the invertibility of matrices with entries*

from a ring, Uchenye Zap. Kishin. Gosud. Univ. **91** (1967), 51-54. (Russian).

[33] B.V. Khvedelidze: *Linear discontinuous boundary value problems, singular integral equations and some of their applications*, Trudy Tbilissk. Mat. Inst. Akad. Nauk Gruz. SSR **23** (1956), 3-190. (Russian).

[34] B.V. Khvedelidze: *The method of Cauchy-type integrals in the discontinuous boundary-value problems of the theory of holomorphic functions of a complex variable*, Itogi Nauki i Tekniki (Sovremennye Problemy Matematiki), **7** (1975), pp. 5-162; English transl.: J. Soviet Math. **1**, No. 3, 309-415.

[35] J.J. Kohn and L. Nirenberg: *An algebra of pseudo-differential operators*, Commun. Pure Appl. Math. **18** (1965), 269-305.

[36] M.A. Krasnosel'skii: *On a theorem of M. Riesz*, Dokl. Akad. Nauk SSSR **131**, No. 2 (1960), 246-248; English transl.: Soviet Math. Dokl. **1** (1960), 229-231.

[37] M.G. Krein: *Integral equations on the half-line with a kernel depending on the difference of the arguments*, Usp. Mat. Nauk **13**, No. 5 (1958), 3-120; English transl.: Amer. Math. Soc. Transl. (2) **22** (1962), 163-288.

[38] N.Ya. Krupnik: *On multidimensional singular integral equations*, Usp. Mat. Nauk **20**, No. 6 (1965), 119-123. (Russian).

[39] N.Ya. Krupnik: *On the normal solvability and the index of singular integral operators*, Uchenye Zap. Kishin. Gos. Univ. **82** (1965), 3-7 . (Russian).

[40] N.Ya. Krupnik: *A Fredholmness criterion for singular integral operators with measurable coefficients*, Soobshcheniya Akad. Nauk Gruz. SSR **80**, No. 3 (1975), 533-536. (Russian).

[41] N.Ya. Krupnik: *Some general problems in the theory of one-dimensional singular integral operators with matrix coefficients*, Mat. Issled. **42** (1976), 91-113. (Russian).

[42] N.Ya. Krupnik: *On singular integral operators with matrix coefficients*, Mat. Issled. **45** (1977), 93-100. (Russian).

[43] N.Ya. Krupnik: *Some consequences of a theorem of Hunt-Muckenhoupt-Wheeden*, Mat. Issled. **47** (1978), 64-70. (Russian).

[44] N.Ya. Krupnik: *A sufficient set of n-dimensional representations of a Banach algebra and the n-symbol*, Funkts. Anal. Prilozhen. **14**, No. 1 (1980), 63-64; English transl.: Funct. Anal. Appl. **14**, No. 1 (1980), 50-52.

[45] N.Ya. Krupnik: *Conditions for the existence of an n-symbol and of a sufficient family of n-dimensional representations of a Banach algebra*, Mat. Issled. **54** (1980), 84-97. (Russian).

[46] N.Ya. Krupnik: *Banach PI-algebras*, in: *Studies on Functional Analysis and*

Differential Equations, Mathematical Sciences, "Shtiintsa", Kishinev, 1981, pp. 54-59. (Russian).

[47] N.Ya. Krupnik: *On a problem in the theory of PI-algebras*, in: *Studies in Modern Algebra and Geometry, Mathematical Sciences*, "Shtiintsa", Kishinev, 1983, pp. 76-79. (Russian).

[48] N.Ya. Krupnik and V.I. Nyaga: *Singular integral operators with a shift along a piecewise-Lyapunov contour*, Izv. Vysch. Uchebn. Zaved. Mat., No. 6 (1975), 60-72; English transl.: Soviet Math. **19**, No. 6 (1975), 49-50.

[49] N.Ya. Krupnik and V.I. Nyaga: *On singular integral operators in weighted ℓ_p spaces*, Mat. Issled. **9** (1974), 206-209. (Russian).

[50] N.Ya. Krupnik and E.P. Polonskii: *The norm of an operator of singular integration*, Funkts. Anal. Prilozhen. **9**, No. 4 (1975), 73-74; English transl.: Funct. Anal. Appl. **9**, No. 4 (1975), 337-339.

[51] N.Ya. Krupnik and I.A. Fel'dman: *On the impossibility of introducing a matrix symbol on some operator algebras*, Mat. Issled. **61** (1981), 75-85 (1981). (Russian).

[52] N.Ya. Krupnik and I.A. Fel'dman: *On the invertibility of certain Fredholm operators*, Izv. Akad. Nauk Mold. SSR, Ser. Fiz. Tekn. Mat. Nauk No. 2 (1982), 8-14. (Russian).

[53] G.S. Litvinchuk: *Boundary Value Problem and Singular Integral Operators with a Shift*, "Nauka", Moscow, 1977 . (Russian).

[54] A.S. Markus and I.A. Fel'dman: *Index of an operator matrix*, Funkts. Anal. Prilozhen. **11**, No. 2 (1977), 83-84; English transl.: Funct. Anal. Appl. **11**, No. 2 (1977), 149-151.

[55] A.S. Markus and I.A. Fel'dman: *On the connection between certain properties of an operator matrix and of its determinant*, Mat. Issled. **54** (1980), 110-120. (Russian).

[56] S.G. Mikhlin: *Composition of singular integrals*, Dokl. Akad. Nauk SSSR **2** (11), No.1 (1936), 3-6. (Russian).

[57] S.G. Mikhlin: *Singular integral equations with two independent variables*, Mat. Sb. **1**(43), No. 4 (1936), 535-550. (Russian).

[58] S.G. Mikhlin: *On a problem in the theory of integral equations*, Dokl. Akad. Nauk SSSR **15**, No. 8 (1937), 429-432. (Russian).

[59] S.G. Mikhlin: *Multidimensional Singular Integrals and Integral Equations*, "Fizmatgiz", Moscow, 1962; English transl.: Pergamon Press, Oxford, 1965.

[60] S.G. Mikhlin and S. Prösdorff: *Singulare Integral-Operatoren*, Akademie Verlag, Berlin (1980).

[61] N.I. Mushkhelishvili: *Singular Integral Equations*, "Nauka", Moscow, 1968; English

transl. of 2nd ed.: P. Nordhoff, Groningen, 1953.

[62] M.A. Naimark: *Normed Rings,* "Nauka", Moscow, 1968; English transl.: Normed Algebras, Wolters-Noordhoff, Groningen, 1972.

[63] S.K. Pichorides: *On the best values of constants in the theorem of M. Riesz, Zygmund and Kolmogorov,* Studia Math. 44, No. 2 (1972), 165-179.

[64] S. Prösdorff: *Einige Klassen Singularer Gleichungen,* Akademie-Verlag Berlin, 1974; English transl.: North Holland, Amsterdam, 1978.

[65] M. Rabindranathan: *On the inversion of Toeplitz operators,* J. Math. Mech. 19 (1969/70), 195-206.

[66] M. Riesz: *Sur les fonctions conjuguees,* Math. Z. 27, No. 2 (1927), 218-244.

[67] L.H. Rowen: *Polynomial Identities in Ring Theory,* Academic Press, New York, 1980.

[68] G.E. Shilov: *Mathematical Analysis (Finite-Dimensional Linear Space),* "Nauka", Moscow, 1969. (Russian).

[69] I.Ya. Shneiberg: *Spectral properties of linear operators in interpolating families of Banach space,* Mat. Issled., IX, No. 2 (1974), 214-229. (Russian).

[70] I.B. Simonenko: *The Riemann boundary value problem for n pairs of functions with measurable coefficients and its application to the study of singular integrals in weighted L_p spaces,* Izv. Akad. Nauk SSSr Ser. Mat. 28, No. 2 (1964), 277-306. (Russian).

[71] I.B. Simonenko: *A new general method of investigating linear operator equations of singular integral type, I.* Izv. Akad. Nauk SSSR, Ser. Mat. 29, No. 3 (1965), 567-586. (Russian).

[72] I.B. Simonenko: *Some general questions in the theory of the Riemann boundary value problem,* Izv. Akad. Nauk SSSR Ser. Mat. 32, No. 5 (1968), 1138-1146; English transl.: Math. USSR, Izv. 2, No. 5 (1968), 1091-1099.

[73] E.M. Stein: *Singular Integrals and Differentiability Properties of Functions,* Princeton Univ. Press, Princeton, N.J., 1970.

[74] E.M. Stein: *Interpolation of linear operators,* Trans. Amer. Math. Soc. 83 (1956), 222-234.

[75] N.L. Vasilevskii and R. Trujllio: *Concerning the theory of Φ-operators in matrix operator algebras,* Mat. Issled. 54 (1980), 3-15. (Russian).

[76] N.B. Vasil'ev: *C^*-algebras with finite-dimensional representations,* Usp. Mat. Nauk 21, No. 1 (1966), 135-154. (Russian).

[77] I.E. Verbitskii and N.Ya. Krupnik: *Exact constants in the theorems of K.I. Babenko and B.V. Khvedelidze on boundedness of singular operators,* Soobshcheniya Akad.

Nauk Gruz. SSR **85**, No. 1 (1977), 21-24. (Russian).

[78] I.E. Verbitskii and N.Ya. Krupnik: *Exact constants in theorems on boundedness of singular operators in weighted spaces, and their applications*, Mat. Issled. **54** (1980), 21-35. (Russian).

[79] A.I. Vol'pert: *The index of boundary value problems for systems of harmonic functions in three independent variables*, Dokl. Akad. Nauk SSSR **133**, No. 1 (1960), 13-15; English transl.: Soviet Math. Dokl. **1** (1960), 791-793.

[80] B.L. van der Waerden: *Modern Algebra, Vols. I, II*, Frederick Ungar, New York, 1949, 1950.

[81] A. Zygmund: *Trigonometric Series, Vols. I, II*, Cambridge Univ. Press, 1959.

[82] R.V. Duduchava: *Discrete Wiener-Hopf equations formed from the Fourier coefficients of piecewise-Wiener functions*, Dokl. Akad. Nauk SSR **208**, No. 1 (1972), 19-22. (Russian).

[83] R.V. Duduchava: *Discrete Wiener-Hopf equations in weighted ℓ_p spaces*, Soobshch. Akad. Nauk Gruz. SSR **67**, 1 (1972).

[84] E.M. Dyn'kin and B.P. Osilenker: *Weighted estimates of singular integral operators and their applications*, Itogi Nauki i Tekhn., Seriya Matem. Analiz, **21** (1983), 42-117; English transl.: J. Soviet Math., **30**, No. 3 (1985), 2094-2154.

[85] N. Jacobson : *PI-Algebras. An Introduction*, Lecture Notes in Math., **441**, Springer-Verlag, Berlin, Heidelberg (1975).

[86] L.V Kantorovich and G.P. Akilov: *Functional Analysis*, Nauka, Moskow (1977); English transl.: 2nd edition, Pergamon Press, Oxford, New York (1982).

[87] N.Ya. Krupnik: *On the quotient norm of a singular integral operator*, Mat. Issled. **10**, No. 2 (1975), 255-263. (Russian).

[88] N.Ya. Krupnik: *Exact constants in Simonenko's theorem on an envelope of a family of operators of local type*, Funkts. Anal. Prilozhen. **20**, No. 2 (1986), 119-120; English transl.: Funct. Anal. Appl. **20**, No. 2 (1986), 144-145.

[89] I.B. Simonenko and Chin' Ngok Min': *The Local Principle in the Theory of One-Dimensional Singular Equations with Piecewise Continuous Coefficients*. Noetherianity, Izd. Rostovskogo Univ. (1986). (In Russian).

SUPPLEMENTARY LIST OF REFERENCES

[12(14)] I. Gohberg and M.G. Krein: *The basic propositions on defect numbers, root numbers, and indices of linear operators*, Usp. Mat. Nauk **12**, No. 2 (1957),

43-118; English transl.: Amer. Math. Soc. Translations (2) **13** (1960), 185-264.

[34(42)]   I. Gohberg: *An application of the theory of normed rings to singular integral equations*, Usp. Mat. Nauk **7** (1952), 149-156. (Russian).

[34(45)]   I. Gohberg and N. Ya. Krupnik: *Systems of singular integral equations in L_p spaces with a weight*, Dokl. Akad. Nauk SSSR **186**, No. 5 (1969), 998-1001; English transl.: Soviet Math. Dokl. **10** (1969), 688-691.

[38(2)]   I. Gohberg: *Some topics in the theory of multidimensional singular integral equations*, Izv. Mold. Akad. Nauk, No. 10 (76) (1960) 39-50. (Russian).

[44(4)]   I. Gohberg and N. Ya. Krupnik: *The symbols of one-dimensional singular integral operators on an open contour*, Dokl. Akad. Nauk SSSR **191** (1970), 12-15; English transl.: Soviet Math. Dokl. 11 (1970), 299-303.

[50(9)]   I. Gohberg and N. Ya. Krupnik: *Introduction to the Theory of One-Dimensional Singular Integral Operators*, Izd. Shtiintsa, Kishinev (1973); German transl.: Birkhäuser Verlag, Basel (1979).

[53(63)]   I. Gohberg and N. Ya. Krupnik: *On the algebra generated by the one-dimensional singular integral operators with piecewise continuous coefficients*, Funkts. Anal. Prilozhen., **4**, No. 3, (1970), 26-36; English transl.: Funct. Anal. Appl. 4, No. 3 (1970), 193-201.